PLANT HARDINESS ZONES

This map, which was developed by the Agricultural Research Service of the U.S. Department of Agriculture, will help you select the most suitable plants for your garden. Every plant included in the "Directory" is given a zone range. The zones 1–11 are based on the average annual minimum temperature. In the zone ranges given, the smaller number indicates the northern-most zone in which a plant can survive the winter and the higher number gives the most southerly area in which it will perform consistently. Bear in mind that factors such as altitude, wind exposure, proximity to water, soil type, snow, night temperature, shade, and the level of water received by a plant may alter a plant's hardiness by as much as two zones.

GRASSES

AND

BAMBOOS

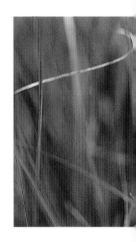

Using form and shape to create visual impact in the garden

GRASSES AND BAMBOOS

NOËL KINGSBURY

photography ANDREA JONES   *consultant* PAUL WHITTAKER

WATSON-GUPTILL PUBLICATIONS / NEW YORK

First published in the United Kingdom
in 2000 by Ryland Peters & Small
Cavendish House
51–55 Mortimer Street
London W1N 7TD

First published in the United States in 2000 by
Watson-Guptill Publications,
a division of BPI Communications, Inc.,
1515 Broadway,
New York, NY 10036

Designer Sailesh Patel
Senior Designer Paul Tilby
Senior Editor Sian Parkhouse
Production Meryl Silbert
Head of Design Gabriella Le Grazie
Publishing Director Anne Ryland

Produced by Sung Fung Offset Binding Co. Ltd.
Printed in China.

ISBN 0-8230-0426-0
Library of Congress Card Number: 99-68050

First printing, 2000
1 2 3 4 5 6 7 8 9 / 07 06 05 04 03 02 01 00

PAGE 1: *Carex comans*
PAGES 2–3: Piet Oudolf's garden (top)
PAGE 4: *Miscanthus sinensis* 'Morning Light'
PAGE 5: *Chimonobambusa tumidissionoda*

CONTENTS

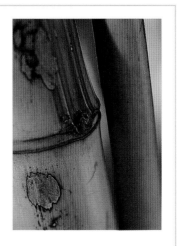

INTRODUCTION

GRASSES, AND THE RELATED SEDGES
AND RUSHES, HAVE MUCH TO OFFER
THE GARDENER: A LONG SEASON OF
ELEGANT GROWTH AND A TOUGHNESS
AND RESILIENCE THAT MAKES THEM
EXCEPTIONALLY USEFUL, ESPECIALLY
FOR SITES EXPOSED TO THE
ELEMENTS. BUT THERE IS MORE THAN
THIS TO GRASSES—THERE IS ALSO AN
INTIMATE RELATIONSHIP WITH HUMAN
HISTORY THAT MAKES THEM
PARTICULARLY IMPORTANT TO US.

OPPOSITE: A variegated bamboo, *Sasaella masamuneana f. albostriata* is
the highlight of this arrangement of plants. The bamboo on the left is
Sasa kurilensis (the dwarf form) and in front is the variegated Japanese
sedge (*Carex morrowii* 'Variegata').

ABOVE: The Japanese timber bamboo (*Phyllostachys bambusoides*
'Castillonis') is one of many bamboos with canes of great beauty.

Grass is one of the most all-pervasive and dominant features of plant life on Planet Earth, and, as lawn, perhaps the most widely grown garden plant. Yet it is only recently that grasses have become widespread in gardens, other than bamboos or lawn grasses. Gardeners and landscape architects the world over are now waking up to the fact that grasses are extraordinarily versatile, resilient, and ornamental plants for a vast number of situations, from tiny backyards to the sweeping grounds of corporate headquarters.

First, we need to look at what exactly grass is, at the role it plays in the world, and at its history, then put it in context. The grass family (Gramineae) is a botanical division containing some 10,000 species, which includes the 1,000 strong subfamily of woody grasses known as bamboos, the majority of which are tropical. But there are many other plants, superficially similar, but which are not botanically classified as grasses because of the different nature of their flowers, the basis of scientific classification of plants. In this book I include sedges (Cyperaceae) and rushes (Juncaceae), which not only look like grasses but behave like them, covering large areas in their growth, a matrix for other plant species. There are also others, known as grasslike plants, that are not grasses and that do not form this matrix-type vegetation, but which are used as grasses in horticulture. In this book, "grasses" is understood to mean true grasses, sedges and rushes, but not "grasslike plants."

It is the matrix quality that develops from the way that many grasses grow that has enabled them to cover so much of the world, forming the vast grasslands that have played such a crucial role in human history. Geologically speaking, grasses are young plants, having evolved only 36 million years ago, whereas the first conifers appeared some 400 million years ago, for example. They have a more efficient method of photosynthesis than other plants, enabling them to grow in the declining levels of carbon dioxide that have been a feature of the history of life on earth (until human pollution at any rate).

One reason why grasses can form such a tight blanket of growth is that their flowers are wind pollinated; they do not have to rely on insects to fertilize them to produce seed, so they can spread rapidly and consistently. These flower heads, turning as the season advances into seed heads, give so many grasses their distinctively attractive character. But there are two separate

LEFT: Often mistakenly called bulrushes, the reedmaces (*Typha* species) are important waterside plants.

ABOVE: Surface features, such as pleating, are an attractive aspect of many bamboos.

OPPOSITE: Pampas grass (*Cortaderia selloana*) is seen through a transparent screen created by another grass.

TOP: Silver grass (*Miscanthus*) varieties
are guarded by pots of the feathertop
(*Pennisetum villosum*).

ABOVE: Leather leaf sedge (*Carex buchananii*)
offers year-round color.

RIGHT: Many of the swith grass (*Panicum*)
species are native to the American prairie.

CENTER RIGHT: Moor grass (*Molinia caerulea*
subsp. *arundinacea*) has fine flower/seed heads.

FAR RIGHT: *Typha laxmanii* is a small reedmace.

habits of growth that have a major effect on how they grow and on the way they can be used in the garden. Turf- or sod-forming grasses produce runners, enabling them to rapidly cover ground without producing seeds. Most species native to cool temperate climates are of this kind, and this is the quality we put to use in lawns. Few such species make good ornamental plants for borders, as they run invasively. In harsher climates, clump-forming grasses dominate, proving their value and stress tolerance in our yards; they form nonspreading clumps or tussocks, which abut each other closely, but leave more space in between for other plant species. They tend to be both more decorative as individual plants and more suitable in the garden. There are also a small number of annual species, including the decorative quaking grass (*Briza maxima*).

The vast majority of grass species are from environments where climatic extremes such as cold or drought prevent tree growth, which is the predominant form of vegetation over most of the world. Before the arrival of humans, grasslands would naturally have been restricted to land above the tree line, coasts, and those regions, usually in the center of continents, with severe seasonal drought.

Human history, even our development as a species, has been intimately tied up with grasses. We evolved in the African savanna, a habitat of scattered trees and stretches of wildlife-rich grassland. It is possible that early humans soon learned that fire destroys trees, or at least stops them from regenerating. It thus assists the formation of grasslands, which could then support large herds of grazing animals, which in turn prevent tree seedlings from re-establishing themselves. The prairies of North America were once limited to seasonally dry areas, but were enlarged by nomadic hunters anxious for more space for the bison on which they depended for food.

The vast areas of natural and semi-natural grassland that cover many of those parts of the globe which have a seasonally dry climate have been an ideal home for the development of nomadic peoples who have followed herds of animals, either wild or domestic. The grasslands of Central Asia have supported such a lifestyle for many thousands of years—a lifestyle that is generally a very productive one, allowing populations to build up. In drought years, though, the threat of famine leads to mass movements of men and animals. The

TOP: La Bambouseraie in southeastern France is an exceptional collection of bamboo.

ABOVE: Eulalia (*Miscanthus sinensis* 'Kleine Fontäne') and hairgrass (*Deschampsia cespitosa* 'Goldtau') feature in a border. BELOW: Bamboo thrives in the foothills of Mount Fuji, Japan.

OPPOSITE: Crookstem bamboo (*Phyllostachys aureosulcata* 'Spectabilis') contrasts with reed grass (*Calamagrostis* x *acutiflora* 'Karl Foerster') and feather grass (*Stipa tenuifolia*) (below).

history of Europe, the Muslim Middle East, and especially China, has been punctuated by periodical outbursts of horse-riding warriors from the grasslands, whose highly mobile way of life and warfare made them formidable opponents for populations dependent upon crops for their livelihood. Some, such as Attila the Hun and Genghis Khan, were savage and destructive; most, however, were more peaceable or more ready to learn from the settled agricultural nations they had invaded, such as the Germanic tribes or the Turks. It is hardly surprising that grass has followed human history. Grass feeds the cattle and other animals that are the mainstay of agricultural economies. All cereals are grasses, too. Fields and pastures have become the norm over vast areas where once the forest held sway.

Pervasive as it is now, the use of grass as an ornamental plant is a relatively late arrival onto the horticultural scene. The enclosed villa gardens of the Romans, the courtyard gardens of Islam, and medieval European monastic gardens had no space for lawns; although medieval knights and their lovers enjoyed the delights of the "flowery mead," this would have been no more than a wildflower meadow within easy reach of the castle walls. Grass made an appearance in Tudor times, for sports such as bowls, but did not become a major part of the garden until the 17th century, when landowners began to lay out extensive grounds. Lawn grass became the foreground to the *allées* and statuary of French master gardener André Le Notre and, later on, rougher cattle-grazed pasture became an essential part of the "English style;" whole landscapes were made into a romanticized arcadia made up of clumps of trees, lakes, and rolling green hills.

Lawn grass was here to stay, but it had to be cut, on a larger scale with sheep or on a smaller scale with teams of workers armed with scythes. A lawn was a status symbol, as owners had to either have enough land to have sheep or be able to employ staff to scythe. It was not until the invention of the mechanical lawnmower in the 19th century that it was possible for more people to have their own lawns, which were soon *de rigueur* for the expanding middle classes of North America and Europe.

While some gardening magazines of the 1880s mention using grasses as ornamental plants, there was little interest in them. Foliage plants may have been popular as house plants, but it seems as if the lack of flower color in grasses condemned them to the margins of the yard. The bolder colors of perennials, subtropical bedding plants, and the increasing number of exuberant flowering shrubs introduced from the East, such as camellias and rhododendrons, attracted far more attention.

Though the Japanese have a long tradition of using grasses in containers, usually accompanying bonsai, interest in ornamental grasses in the West really got underway in postwar Germany, when nurseryman, designer, and writer Karl Foerster began to use them in naturalistic compositions alongside perennials. He recognized their value as plants with a long season of interest, even through harsh winter weather. Richard Hansen, who was also German, worked on a study of perennials in public places, using grasses as accompaniments to conventional flowering plants. Gardeners in other countries then started to use them, with particular interest in the United States. Today the leading lights in the use of grasses are Wolfgang Oehme (who trained in Germany in the 1950s) and James

van Sweden in the United States, and Piet Oudolf in the Netherlands. Oehme and van Sweden use them in concentrated blocks, inspired by the prairies of the Midwest, while Oudolf creates borders using a dramatic blend of grasses and perennials.

Bamboos are naturally plants of moist temperate forests. They characteristically form understory vegetation, especially in the tropics (above a certain altitude) and in areas with a warm humid summer. Though they are usually associated with the Far East, there are in fact many species native to Africa and to South America. Some species

BELOW LEFT: Silver hairgrass (*Koeleria* species) grows alongside pink knapweed (*Centaurea dealbata*), demonstrating how easily and effectively grasses and bamboo blend with perennials in mixed plantings.

BELOW RIGHT: The fescues are tightly clump-forming grasses, with sharp very narrow evergreen leaves. Maire's fescue (*Festuca mairei*) has a distinct arching style and is suitable for planting in dry places.

cover large areas of ground with impenetrable growth; others climb voraciously, smothering whole trees. They have been in cultivation in the East for centuries, where their elegance has long been recognized, and their utility no doubt for much longer. Bamboo canes combine lightness with strength, making them ideal constructional material; perhaps this is why they are seen as having an important value as the symbol of the relationship between human life and nature.

Bamboo is a favorite subject in Chinese and Japanese art, and has long been popular in gardens in both countries; particularly effective use of bamboo is made in Japanese courtyard and temple gardens. Everything in Far Eastern gardens is highly symbolic and representational, with bamboo symbolizing vigor and its canes used to represent forest. Modern Japanese gardens often include nontraditional plants such as yucca and phormium alongside bamboo; they are chosen so their foliage presents contrast.

Bamboos were introduced to the West during the 19th century, with particular growth in interest after trade with Japan was established. Many of the first plantings were on large estates, the bamboos being gifts from silk traders to the landowner. Certain extremely invasive species did rather too well in some gardens and the whole group acquired rather a bad reputation. For many years, only a limited number of species have been available commercially, which has also limited their popularity. The more widespread availability of a wider range of species has done much to popularize them, as has a network of enthusiasts throughout the U.S., Britain, and France.

GRASSES HAVE A LIGHTNESS OF
TOUCH UNEQUALED BY ANY OTHER
GROUP OF PLANTS. IN SEEKING TO
UNDERSTAND HOW THEY CAN BE
USED IN THE GARDEN, IT HELPS
TO UNDERSTAND THE QUALITIES THAT
MAKE THEM SPECIAL, ESPECIALLY
SINCE THESE QUALITIES MAY BE
VERY DIFFERENT FROM THOSE OF
THE MORE CONVENTIONAL GARDEN
PLANTS WE ARE FAMILIAR WITH
IN OUR YARDS.

QUALITIES

OPPOSITE: A sedge (*Carex* species) lurks in a border, awaiting
its season of glory when everything around it has died down.

ABOVE: Snowy woodrush (*Luzula nivea*), a shade-tolerant
evergreen, has a brief period of flower in early summer.

FORM

This is perhaps the greatest contribution grasses have to make to our yards. It is an undervalued aspect of plants; far too many people do not see beyond color. And yet even the most colorful yard can be dull if no attention is paid to the plant forms. It is instructive to take a black-and-white picture of a garden or border, and then see what it has to offer; many colorful plantings are reduced to a jumble of indistinct gray. For much of the year,

there are few flowers in evidence anyway. Rather than give up on the garden altogether, as many do, or just relying on the same few boring evergreens, consider how it might be possible to use a variety of plants with distinctive shapes.

Some grasses, the big and bold ones, can make a definite impression on a large scale. They might be the only grasses in a gard or planting plan and still be a major presence. A single clump of bamboo or pampas grass (*Cortaderia selloana*) is capable of dominating a large area, 365 days of the year. Equally distinct but smaller grasses, reed grass (*Calamagrostis* x *acutiflora* 'Stricta'), for example, can command a smaller area simply by having a growth pattern that is totally distinct from plants around it.

Some gardeners like the stimulating effect of combining lots of different shapes, rather like some people enjoy bringing lots of different colors together, whereas others feel the effect created is too fussy or over stimulating. The former group may well want to mix several different grasses or at least to involve several within a border, counterpoising a variety of different shapes. The latter will probably want to stick to just one or two, to provide structure to a collection of more amorphous plants.

Opinion is divided about combining different grasses, even among enthusiasts. Some feel that they like to keep them separate; others that it is fine to mix different grass varieties. Personally I like to mix them, but to keep it simple, using multiples of two or three species, each with a distinct habit. One of my favorite plantings features the bolt upright reed grass (*Calamagrostis* x *acutiflora* 'Karl Foerster') surrounded by much lower-growing mound-forming hardy geraniums and a low clump-forming brown sedge (one of the *Carex* species). They contrast brilliantly, but do not compete.

Some grasses are deciduous; they live with us for only part of the year, such as the big star burst of moor grass (*Molinia*); others, such as sedge (*Carex*), are evergreen and a permanent feature. Bamboo plays a role in the yard not unlike that of evergreen shrubs—they are woody and their structure offers bulk and permanence. One of the most exciting aspects of gardening with grasses is the seasonal dynamism of plant forms: the constant change of deciduous species' growth, death, and renewal against a background of permanence.

OPPOSITE: The stiffly upright form of reed grass (*Calamagrostis* x *acutiflora* 'Karl Foerster') makes it one of the most valuable of all garden grasses.

LEFT: A silver grass (*Miscanthus*) cultivar (left), a bamboo, and a sedge (*Carex*) illustrate the variety of form found among grasses and allied plants.

ABOVE: Reed grass (*Calamagrostis*) is one of many grass species that brings life to the late summer garden.

BELOW: *Miscanthus sinensis* 'Sarabande,' a reliabe cultivar of Japanese silver grass, has elegant leaves.

RIGHT: These almost treelike canes are those of the bamboo *Phyllostachys edulis*.

FAR RIGHT: The young growth of bamboo has a much stiffer upright habit than the mature stems. This is yellow-stemmed crookstem bamboo (*Phyllostachys aureosulcata* 'Aureocaulis').

UPRIGHTS

The majority of garden plants do not have good form; they are often rather shapeless and indistinct. The use of a few strong shapes can transform a garden – and nothing so much as a few good vertical lines, especially in flat areas, or wide open gardens. Grasses provide some of the best verticals you can grow, and they are a lot quicker to establish than classic vertical plants such as a palm tree or an Italian cypress.

Close up, tall bamboos, such as the crookstem bamboo (*Phyllostachys aurea*), can form wonderful uprights, and they are made great use of in Japanese courtyards, but on the whole bamboos arch too much in temperate climates to stress the vertical. Very tall grasses, such as the giant reed (*Arundo donax*) or silver banner grass (*Miscanthus sacchariflorus*), are impressive but leafy, which detracts from the bold skywards sweep. Any upright grass—pampas grass (*Cortaderia*) or silver grass (*Miscanthus*), for example— makes a good display of vertical stems for up to six months, especially with the contrast in color and line between the stems of the flower/seed heads and the clumps of leaves.

There are a few grasses that are so superbly upright that others are put in the shade, notably the reed grasses *Calamagrostis* x *acutiflora* 'Stricta' and *C.* x *a.* 'Karl Foerster,' which produce bolt-upright flower/seed heads that will last for at least nines months of of the year, from midsummer through to early spring. They are completely windproof, being resistant to hurricane-force winds according to James van Sweden, who uses them in coastal gardens on the Atlantic seaboard. They have small feathery tops on the tall canes, which are very different in effect from the plumelike, generally one-sided heads of pampas grass (*Cortaderia*) and silver grass (*Miscanthus*).

Strong verticals in the yard need careful placing and repay considerable thought over grouping. Those that have bulk as well as an upright posture tend to dominate their surroundings, one reason perhaps why pampas grass (*Cortaderia*) has become unpopular in many quarters. Those that form less of a clump or are more graceful in build are easier to mix in with others. It is common sense not to have strong vertical lines crowded by other plants at the same height; there seems little doubt that a really good vertical should stand well clear of any surrounding plants, which should be much lower growing.

Having a few clear verticals arising out of other plants can be very dramatic indeed, but some people prefer the effect of a multiple planting. Personally, I feel the reed grasses (*Calamagrostis* species) already mentioned look better planted as single plants that will then grow into small clumps over the years, rather than grown as several massed plants. When they are grown in a large group, I feel that part of the essential qualities of the individual plants is lost. However, I know that certain of my colleagues prefer the sight of massed ranks of upright stems, recalling a reed bed, and grown this way they certainly act as a good back-drop to lower-growing perennials or groundcover plants. And tall grasses can make good temporary summer screening or shelter.

ELEGANCE

It is difficult to describe elegance, although we all know it when we see it. Bamboos for many gardeners are the epitome of elegance; it has something to do with the way the stems grow up and then arch over, with the leaves held outward and slightly weeping. The shadows created by bamboo foliage on hard surfaces also have a particular magic. This arching quality seems to be at the heart of the elegance of many grasses, too. Even something as robust as the weeping sedge (*Carex pendula*) has extremely elegant flowers with long catkins hanging from a long, gently arching stem.

The combination of arching stem and pendent head is key to the appeal of many grasses. The comparatively large flower/seed heads of spangle grass (*Chasmanthium latifolium*), or of the annual *Briza major*, hanging from threadlike stems, are good examples—the latter is known as "quaking grass" since the heads seem constantly in motion. The plume heads of so many grasses recall the wild reed: they all tend to blow in one direction, with the stems all having exactly the same curvature—a powerful image. Silver grass (*Miscanthus*) on a windy day in winter is a wonderful example. Many of the melic grasses (*Melica*) have similar qualities, but on a much more intimate scale.

Elegance and grace are qualities that nearly all the feather grasses (*Stipa*) have, whether it is the immensely long awns of *Stipa pennata*, or the tall stems of giant feather grass (*S. gigantea*) with their open panicles of narrow oatlike heads. It is a quality that draws attention through its subtlety, and giant feather grass is certainly subtle. It is undeniably a huge grass, up to six feet tall and wide in flower, yet it never dominates because its stems are so narrow and so far flung it is effectively transparent.

It is the leaves that are the source of grace in some grasses—those with particularly finely shaped leaves, for example, such as many of the feather grasses. Even if the leaves themselves are rather coarse in appearance, the way they are arranged around the stem can be especially pleasing to the eye, as in frost grass (*Spodiopogon sibiricus*).

Elegance is a virtue that needs space to be appreciated. Being such a subtle quality it is easily overwhelmed. Elegant plants should never be cramped, and they should be combined with plants or surroundings of quiet beauty. Specimen planting, surrounded by lawn or groundcover, suits some of the larger species, while careful placing is all that others need, so they are not hidden, but are easily seen and thus appreciated.

The traditional Japanese garden, or courtyard, is the perfect place to appreciate elegance and perfection. Fine plants are grown as single specimens, or perhaps a group of several individuals of the same species are planted together, often counterpoised by carefully chosen gravel, stones, or decking. There is always a contrast of hard and soft textures, straight and curved lines. This is the perfect way to appreciate bamboos, the grace and refinement of practically all parts of all species rewarding close study.

ABOVE FAR LEFT: Tough and sometimes invasive because of self-seeding, weeping sedge (*Carex pendula*) is nevertheless valuable for bringing its unusual quality of elegance to dry shade.

ABOVE LEFT: Remains of frost grass (*Spodiopogon sibiricus*) seed heads rise above the foliage in autumn.

ABOVE: Giant feather grass (*Stipa gigantea*) is one of the most widely grown ornamental grasses.

CLUMPS

The clump is the growth habit of a very large number of grass species, but here I wish to concentrate on those that form a relatively low clump. Such dumpy plants are quite the opposite of the graceful ones we have just considered, but this does not mean to say that they have no value in the yard. On the contrary, a yard needs large numbers of quietly attractive plants to give the star performers space and to let them be seen to best advantage. Many garden plants, in particular the more dramatic ones and the vast majority of later-flowering perennials, are upright growing in growth pattern, and a lot of them tend to develop unattractive or bare stems at their bases. They need lower plants around them to hide their stems, cover the ground and fill in around them, especially at the front of a bed.

In the wild, some clump-forming grasses form very large, distinct tussocks, a nightmare to walk across. This natural look can be captured in the yard if tight clump-forming species are grown in proximity, and perhaps combined with naturally rounded hump-forming dwarf shrubs such as hebes and lavenders. Such shapes are not eye-catching, but they have a pleasing harmonious quality.

Species of sedge (*Carex*) include many of the best clump-formers, with their tight masses of neat leaves, very often interestingly covered. Most have arching leaves that start off by growing upward and then radiate out or arch over, effectively filling space at ground level, a habit often useful in containers or raised beds. Some, like the plantain-leaved sedge (*Carex plantaginea*), have fewer, broader leaves, which contrast well with other species. The woodrushes (*Luzula*), have a similar form to many of the *Carex* species and a slightly spreading habit that makes them good groundcover plants.

Grass clumps vary greatly in size: some fescues (species of *Festuca*) form very small tufts, mainly worth growing for their distinctive color. Plants with such a growth form are useful for filling in among heathers, other dwarf shrubs, and low-growing rockery plants. At the other end of the scale, pampas grass (*Cortaderia selloana* 'Aureolineata') forms really quite large clumps of distinctly yellow-lined leaves. They are big enough to be a valid part of shrub plantings, and they offer a good contrast in form.

While it is the foliage of these plants that earns them a place in the yard, it is in fact the flowers that develop from some grasses, such as hair grass (*Deschampsia*), that gives them their unique character. Like many grasses, they produce multiple panicles of spaced-out flowers/seed cases that form a softly textured cloudlike mass above and around the foliage, which is very useful for attractively filling in space at the front of the border around leggy late-flowering perennials.

OPPOSITE: The distinctive clumps of Maire's fescue (*Festuca mairei*) form part of a dry landscape planting.

ABOVE: In mild areas feather grass (*Stipa arundinacea*) will stay green for much of the winter, turning to russet. Its form is always attractive.

LEFT: Sheep's fescue (*Festuca amethystina*), normally blue-gray, takes on its autumn mode.

TEXTURE

Plants can look soft, such as a grass that forms a cloudlike head of masses of delicate flowers, or hard, such as one that has broad, glossy leaves. Texture, like form, is one of those qualities that gives extra interest to a garden, and it is particularly useful at the end of the season or in winter, when there is not so much else to look at.

Texture also plays an important part in the way we look at a garden; soft-textured plants appear matte to the eye, absorbing light, which has the effect of making them recede, looking further away. Hard textures stand out; a glossy surface, or the pattern made by large or striking leaves, draws attention to itself, and thus appears closer than it really is. It thus stands to reason that a garden or border is going to look much smaller if a bold, especially a big and bold, plant is placed at the far end.

Plants form quite different patterns with their leaves. For example, *Luzula nivea* 'Marginata,' forms rough rosettes of radiating broad leaves in contrast to its relative the snowy woodrush (*L. nivea*), whose leaves are softer and narrower. Both make good groundcover in shade; the former looks quite striking, almost tropical in appearance, while the latter is more grasslike and consequently less striking, except for the brief period of its flowering season when its soft creamy heads make a dramatic display.

Some grasses tend to look a bit untidy after the wind has blown them about, whereas others maintain a remarkably well-tended look. A good example of the latter is Hakone grass (*Hakonechloa macra*), whose leaves look surprisingly neat, as if the fairies comb them every night, sweeping out from the center of the clump.

The different textures bamboos develop is one of their distinguishing characteristics. Those with small leaves, such as *Fargesia nitida*, look quite different from those like species of *Indocalamus,* which have much larger leaves, or those with long, narrow leaves like *Pleioblastus linearis*. Species with leaves that hang right down to the ground, such as *Yushania anceps*, stand out too. Not everyone likes to mix bamboo species, but for those who do, juxtaposing different textures is all part of the fun.

Bamboos are marvelous plants for looking at close to—the texture of the tall canes can be smooth and glossy, or they can be covered in a hazy bloom. Some have glossier leaves than others; some have striations on the leaves, others not. The emerging stems on mature plants in spring (known as culms) are also very attractive. It is these characteristics that make them such rewarding plants to have in a small space or somewhere where you frequently walk past them.

COLOR

However attractive the forms and textures of so many grasses, I suspect that a large number of the people who have bought them for their yards have done so for their foliage color. For those unused to dealing with them, this is understandable. Most of us are more familiar with and thus more confident at dealing with color in the garden than form, and it tends to be seen as a more important factor.

Grass foliage colors can never compete with flower colors, but why should they? Flower colors are so intense, compared to most foliage colors, that the eye can only take in so many of them. Most of us prefer flower colors to be in a certain proportion to less intense foliage colors; although lovers of massed ranks of bedding annuals will beg to differ. Green is, of course, the most important foliage color, and one that is relaxing to look at, the perfect backdrop for flower color. But some plants have brown, bronze, or purple tones in their foliage, some even have pink and red. Grasses and sedges are particularly strong on these colors. Silver, gray, and blue foliage seem to enjoy periodic swings and falls in garden fashion, but always sell well in garden centers, whatever the current vogue. They provide variation and look particularly effective with many flower colors. Finally, there are all those plants with variegated foliage, generally classified as having either white/cream/silver variegation or gold/yellow.

The great advantage of foliage colors over flowers is their much longer season, especially if they are evergreen. Flowers come and go, but foliage is there from spring to autumn at least, a boon in small yards where every plant has to earn its keep. Such a long season provides continuity, and can be used to establish a link from one season to the next. It is this satisfyingly long season, and the wide variety of color, that has encouraged so many gardeners to turn to grasses.

TOP LEFT: *Achnatherum brachytricha* is an effective and relatively compact grass for late summer and autumn.

TOP CENTER: Tussock grass (*Chionochloa rubra*) contrasts with the blue-gray foliage of a cardoon (*Cynara cardunculus*).

TOP RIGHT: *Milium effusum* 'Aureum' is known as Bowles' golden grass after a famous English gardener. It is useful for early summer color, especially in light shade.

LEFT: Sunlight catches the variegated leaves of pampas grass (*Cortaderia selloana* 'Silver Stripe').

BELOW: Hairgrass (*Deschampsia cespitosa*) is a common plant of light woodland, which makes it eminently suitable for light shade in the yard.

RIGHT: The fresh green of the autumn moor grass (*Sesleria autumnalis*) is useful for edging.

FAR RIGHT: Feather grass (*Stipa arundinacea*) has a long season of autumn color.

GREENS

Green is such a dominant color in the garden that it is rarely discussed. Yet there are so many different greens. I particularly like the fresh, slightly lemon green of melic grass (*Melica altissima* 'Atropurpurea'), which is one of those greens that looks fresh all the way through the year. Then there are the darker shades of green of the hairgrass (*Deschampsia cespitosa*) and of many bamboos. In between there are endless variations, many of which are only really appreciated when the plants are seen together.

When selecting plants or trying to place them in the yard, try to hold flowers against leaves and see which combinations look effective. Some greens definitely seem to intensify certain colors and dull others. A good example is the combination of scarlet with a clear medium green, which seems to electrify both.

ADDING RED TO GREEN

Grasses, and perhaps even more, sedges, are remarkable for the wide range of foliage colors that result from the addition of red to green: browns and bronzes. The most outstanding is the Japanese blood grass (*Imperata cylindrica*) with its wonderful red leaves, at their most dramatic when backlit. While there are no grasses, yet available commercially anyway, with purple foliage, there are plenty in the bronze-to-brown spectrum. These all look remarkably attractive together, as the eye seems to appreciate a cavalcade of different and subtle variations on the same theme.

Of the bronze/brown leaved grasses, the finest, I think, is feather grass (*Stipa arundinacea*), which has the added advantage of gradually developing its color over time. During the growing season, its arching leaves are dark olive green, turning golden-bronze in autumn, a tone that deepens through the winter, before dying to brown in the coldest weather. An evergreen, and less dynamic, version of these colors is offered by several New Zealand sedges that are available; orange sedge (*Carex*

testacea) and *C. dipsacea*, for example, have leaves that are a subtle blend of yellowish-green and greenish-yellow, appreciated especially in the middle of winter.

Of the browns, the bronze form of New Zealand sedge (*Carex comans*) is often derided for looking dead. Yet no dead plant could have such a deep luster to its fine bronze-brown leaves. But this color is definitely one to be appreciated in context, as a contrast to green, or as part of a spectrum of similar colors. It works well as a backdrop for yellow flowers; I grow it with late-summer-flowering *Rudbeckia fulgida* var. *deamii*.

The great importance of the sedges is the opportunity they offer of using ever-greens in the border. Traditionally, evergreens have been limited to trees and shrubs, and borders have remained bare and brown through winter. Evergreen sedges in traditional dark green such as the weeping sedge (*Carex pendula*) offer the opportunity to add this color to the border for the winter, while the colored-leaved species offer the more exciting possibility of dots and splashes of bronze and gold among the dormant perennials of winter. When spring starts, daffodils and narcissi find an echo of their flower colors among the sedges—this is important, for I find early spring bulbs seem to flower amid such bare surroundings, often lacking context for their flowers.

Given their ability to cope with poor soils and exposed conditions, perhaps the best use for all the bronze- and brown-leaved sedges is in combination with other plants that thrive is similar conditions: heathers, low-growing rhododendrons, and dwarf conifers. Their colors mix well with the colors of heather and conifer foliage, and they offer a striking contrast in form to the mound-forming heathers and spreading conifers.

ABOVE: Staying more or less the same color all year round, the sedge *Carex dipsacea* is another valuable source of winter interest.

SILVER AND GRAY

Some plants have a dense pattern of minute hairs on their leaves that gives them the appearance of having silver, gray, or sometimes even blue-gray foliage. Most are from environments that regularly experience drought; the hairs help prevent water loss. They are of great value in the garden for the change they offer from green and the way they intensify colors such as pink, blue, and purple—in fact they work well with all colors except for yellow, which they tend to muddy. Gray and silver foliage mix well together, too, making it possible to blend a variety of grassy and non-grassy species.

The deepest of these tones is that of wild rye (*Elymus magellanicus*), which is definitely blue, but it can be a short-lived plant, especially on moist soils. Other species of wild rye are good, too, but tend to spread invasively. Perhaps the best all-

round medium-sized silver grass is blue oat grass (*Helictotrichon sempervirens*), which combines reliability with good color. Plants of this size are effective as accent plants dotted around borders, and are useful for creating continuity when scattered in this way. They look good alongside gray-leaved plants from similar dry environments such

TOP: Blue oat grass (*Helictotrichon sempervirens*) is one of the most reliable blue grasses.

ABOVE: Sheeps' fescue (*Festuca amethystina*) is one of many low-growing grasses whose foliage offers the gardener a wide range of blue, gray, and glaucous tones.

as low-growing rock roses (*Cistus*) and lavender, their habit a contrast to the mound forms that so many dry climate or Mediterranean climate dwarf shrubs tend to take.

Large blue or gray grasses are very few and far between, although the prairie native Indian grass (*Sorghastrum avenaceum* 'Sioux Blue') grows to over three feet, and is thus potentially suitable for mixing with larger perennials. On a different scale are the silver-blue forms of blue fecue (*Festuca glauca*), of which several are now available. Their color is good, and they are easy to grow; they thrive on poor soils, so are useful companions for heathers, dwarf conifers, and the other grasses that flourish on these soils. Their small size may appear to limit their usefulness, but not if they are used *en masse* as groundcover, where they are a very effective contrast with larger plants like heathers. Their clump-forming habit does mean that there tend to be gaps between the plants, but I find this problem can be overcome if they are combined with species of *Acaena*, low-growing groundcover plants with tiny leaves in silver, gray, and brown shades, or other similar rockery-type plants. The result is a groundcovering tapestry of subtly different shades. Such groundcover is effective alongside paving stones or filling the gaps between stepping stones. Creeping thymes and lawn chamomile could be combined with these plants, too.

TOP: Wild rye (*Elymus magellanicus*) has the best blue of any grass. Although short-lived, it is easily raised from seed.

RIGHT: Fescue (*Festuca*) are tolerant of sandy and limestone soils and exposure.

VARIEGATION

Variegation has long been prized in plants, partly for its novelty value, but mostly for the additional options it provides for creating plant compositions based on foliage color. We have already seen how valuable grasses are with leaves in colors other than green. The value of variegation, however, lies not just with the leaf color but with the pattern that the variegation takes; this may not be of great importance on a larger scale, but can be a factor that adds greatly to a plant's interest in small-scale or intimate schemes. The streaks and banding of silver or gold alongside the green of most variegated grasses is often worthy of this close attention. Sometimes, though, a variation

occurs that results in a grass having yellow-flushed foliage rather than true variegation. One such is Bowles' golden grass (*Milium effusum* 'Aureum'), a species tolerant of shade, where its light golden color is especially welcome.

Most grasses with variegation are gold rather than silver, and most have it as an edging to the leaf. A form of pampas grass (*Cortaderia selloana* 'Aureolineata') is the largest of these. Its leaves are minutely edged with yellow, creating a neat mound of foliage that seems to glow soft gold, a great improvement on the coarse leaves of the normal form. Hakone grass (*Hakonechloa macra* 'Aureola') is another fine golden variegated grass, but it grows to only a foot. *Pleioblastus auricomus* is a golden bamboo. These truly golden species combine well with flowers in "hot" colors while most other yellow variegated grasses tend to be paler, which makes them less outstanding in themselves but more suitable for combining with a wide variety of other colors.

TOP: "The first grass gardeners are given—and the first they throw away" is how a colleague describes ribbon grass (*Phalaris arundinacea*). Despite its aggressive growth, the variety 'Feesey' is an attractive plant.

ABOVE: Porcupine grass (*Miscanthus sinensis* 'Strictus') has a unique pattern of variegation.

Cream variegation is perhaps the most common form among grasses; Japanese sedge (*Carex morrowii* 'Evergold') is one of the best, and it is useful for lightly shaded plantings and combining with groundcover plants, especially those with broad foliage like bergenias and hostas. In contrast to the fine leaves of this species, *Carex siderosticha* 'Variegata' has very broad leaves, making it a good contrast with ferns.

White, rather than cream, variegation, is, like the true gold, not so common. Clean white variegation is invaluable for mixing with strong colors, to calm them down and offer contrast. The finest is that of the giant reed (*Arundo donax* var. *versicolor*). It is not very hardy, but it is so beautiful and so exotic looking that it may be worth bringing under cover in winter. *Calamagrostis* x *acutiflora* 'Overdam' has white variegation and the splendidly upright habit of its parent the reed grass (*C.* x *a.* 'Karl Foerster'), making it useful for combining with lower-growing perennials with vivid flowers such as geranium or penstemon species.

Nearly all grass variegation happens along the length of the leaf; in only a few similar varieties does it occur across the leaf, to form bands: zebra grass (*Miscanthus sinensis* 'Zebrinus') and porpcupine grass (*M.s.* 'Strictus'), for example. The plants are striking for their novelty as well as being yellow enough to consort well with hot colors.

TOP LEFT: *Pleioblastus shibuyanus* 'Tsuboi' is a highly effective variegated bamboo. The dark green of *Indocalamus tessellatus* forms a strong contrast.

TOP RIGHT: Yellow-leaved Bowles' golden grass (*Milium effusum* 'Aureum') grows at the waterside.

ABOVE: Japanese sedge (*Carex morrowii* 'Evergold') makes a good container plant because of its neat habit and evergreen foliage. It also thrives in light shade, so it is a versatile addition to a yard.

DRAMA

Plants can impress us through their color, their form, or through other qualities. Of these, drama is one of the most important to the gardener, and bamboos and grasses have plenty to offer here. But what makes a plant dramatic?

The unexpected is a major component: plants with very unusual foliage or shape that causes surprise when first seen. Size, too, often surprises: suddenly coming across a plant that is larger than one expects. We tend to think of grasses as being relatively small plants, so extra-large ones tend to stand out. Large foliage, whether it takes the form of very broad or very long leaves or tall stems, can surprise, too, especially since it tends to be associated with tropical species, and can astonish when seen flourishing in a cooler climate. Foliage shape in itself can be a component of drama—leaf shapes that are unusual stand out as effectively and for the same reasons as unusual architecture or startlingly designed clothes.

Large grasses are some of the most important plants for the gardener who wants to introduce drama into the yard. They are a fantastic antidote to the "twee" cottage style and safely soothing planting plans that are so heavily promoted. Big, almost oversize, plants and striking leaves provide a feel that is contemporary and dashingly forward looking. US designer James van Sweden told to me that "small plants only make a small garden smaller" as he showed me his tiny back garden in downtown Washington, DC, dominated by the huge sugarcanelike stems of giant miscanthus (M. floridulus).

Silver grasses (Miscanthus) include some of the most effectively dramatic grasses. It is not just the large size of some of them, but the elegant sweep of their seed heads above solid clumps of dark leaves that contribute to this quality. Other grasses create drama simply through sheer size, such as the gray-leaved but scruffy giant reed (Arundo donax). This, and larger species of silver grass (Miscanthus), do best on moist soils, as do many plants with large foliage. They thrive alongside, and look good with, perennials with broad leaves such as ornamental rhubarb (Rheum) or giant rhubarb (Gunnera). Combining such plants can create a very exotic impression; those gardeners who wish to conjure up the tropics or hint at the primeval should turn first to this plant selection.

Bamboos are also a good source of dramatic plants. Those that grow tall, such as Semiarundinaria fastuosa, are particularly striking, as are those that develop thick canes like Phyllostachys violascens. In Japan, bamboo with thick canes is frequently included in courtyard plantings, the canes rising out of gravel-covered earth, to grow upwards, often with the foliage up and out of sight, the whole presenting a planting dramatic in its simplicity, but also hinting at a forest of tree trunks. Large-leaved bamboos, such as species of Indocalamus and Sasa, are also particularly dramatic and exotic in effect.

ABOVE FAR LEFT: Porcupine grass (Miscanthus sinensis 'Strictus') is strongly banded.

ABOVE LEFT: Miscanthus sinensis 'Africa' is one of many dramatic varieties of silver grass bred by German nurseryman Ernst Pagels.

FAR LEFT: Miscanthus sinensis 'Silberturm' is the tallest of the cultivars.

LEFT: Giant miscanthus (M. floridulus) is grown for its impressive foliage.

RIGHT: Miscanthus sinensis 'Ghana' is a new cultivar with rich coloring.

ABOVE: Elegant flowers and habit and good autumn color—*Miscanthus sinensis* 'Flamingo' is first-class.

BELOW: Reed grass (*Calamagrostis* x *acutiflora* 'Karl Foerster'), pampas grass (*Cortaderia selloana*) and Japanese silver grass (*Miscanthus sinensis*) form the framework for a late summer border.

RIGHT: The smooth-sheaved bamboo *Phyllostachys vivax* 'Aureocaulis' looks particularly poetic in the snow.

BELOW RIGHT: *Miscanthus sinensis* 'Pünktchen' combines variegated stripes with reliable autumn flowering.

SOUND AND MOVEMENT

Walking through the countryside can be one of the best ways of appreciating some of the special features of grasses. Fields of long summer grass, or fields of wheat and barley, are in a state of constant movement in the slightest breeze; waves of motion cross the grass, as the flower heads bow to the wind and then spring back upright, before bowing again to the next gust. All the time, there is the sound of wind swishing its way through the grass. Larger grasses, like fields of corn, bow less to the wind, but make more noise, as the large leaves rustle constantly against each other. Reed beds particularly are brought to life by the wind: the flower or seed heads all point in one direction, blown to one side by the prevailing wind, bending and bouncing back *en masse*. Their large and broad leaves make an especially distinctive sound.

Bamboos, with their large hard leaves and hollow stems, make a particularly dramatic sound as they move in the wind, the leaves slapping against each other, and the canes rattling. The larger the bamboo, the greater the sound. I shall never forget being in the mountain rain forest behind Rio de Janeiro beneath a vast bamboo in a storm. Huge canes banged into each other, others rubbed and scraped, making a sound that I can only liken to the creaking of a sailing ship, all against a backdrop of rustling leaves.

Nothing quite like this can be achieved in the temperate garden, but sound and movement are still features that give grasses a special place in the yard. Many grasses move in the slightest breeze, situations where other perennials stand stock still. The quaking grasses, the annual *Briza maxima* and the perennial *B.media*, with their

BELOW LEFT: Always in motion, the annual quaking grass (*Briza maxima*) flowers only a few months from seed.

BELOW RIGHT: The banding on its stems makes the fishpole bamboo (*Phyllostachys aurea* 'Koi') a fine form.

BELOW CENTER RIGHT: A detail of a flower head of *Miscanthus sinensis* 'Ghana' shows its delicate feathering.

relatively heavy flower and seed heads hanging from light wiry stems, are particularly susceptible to air movement, as their common name indicates. Especially beautiful are those species, like silver grass (*Miscanthus*), that have one-sided flower and seed heads, much like wild reeds; all bend to one side and move in unison as the wind blows through them. Planting such varieties in a situation where they are outlined against the sky and exposed to the wind assures you of appreciating their special qualities.

Bamboos make good plants for courtyards and other confined spaces, partly because of their qualities of sound and movement. The absence of the sights and sounds of nature is one of the reasons why urban environments can be alienating to many people. Bamboos, with their response to the slightest breath of wind, are a very effective way of introducing such natural feelings and events into built-up places.

If these qualities appeal, you can be a connoisseur of grass sounds. Each species really does make a different sound as it moves in the wind, depending upon the size and number of leaves and the height and distribution of the seed heads.

RIGHT: Giant feather grass (*Stipa gigantea*) is nearly always in motion.

BELOW RIGHT: *Miscanthus sinensis* 'Kaskade'—its name evokes the quantities of hanging flowers that this cultivar of Japanese silver grass produces, which sometimes weigh down its stems.

CHANGE AND CONTINUITY

Nature is in a constant state of flux and change, or growth and death. For many people, this is an important part of the enjoyment of the garden, having spring flowers to look forward to, or autumn fruits. But it also creates certain problems: practical ones such as the need to clear away dead material and esthetic ones such as the fact that there are times of the year when there is little to appreciate. It is not surprising that evergreens are popular—they provide continuity through the year, as well as always looking good. But the fact that they look pretty much the same, all year round, is a disadvantage; something that is always the same can be boring.

Grasses offer the gardener opportunities to explore change and continuity in ways that are highly advantageous. Grasses may flower, but their flowers are, in nearly all cases, pretty much identical to their seed heads, and most of us fail to notice when the former die and turn into the latter. Conventional flowers can be nine-day wonders, but this is not the case with grasses: their flower/seed head interest often last for months. Perhaps the briefest period of interest is that of snowy woodrush (*Luzula nivea*), with about ten days of cream flowers in early summer, or Siberian melic (*Melica altissima* 'Atropurpurea'), which has wonderful dark purple flowers a little later, but whose color is lost after the flowers die. Many, though, flower in summer, develop seed heads, and can still look attractive in winter, such as the plumelike heads of silver grass (*Miscanthus*). Reed grass (*Calamagrostis* x *acutiflora* 'Karl Foerster') has perhaps the longest season; flowering in early summer, its seed heads still make a dramatic feature in the yard the following spring.

While evergreens provide the most obvious sense of continuity in the yard, the long season of many grasses contributes continuity, too, but one that also includes a feeling for the natural rhythm of the seasons. The interesting and rather convenient thing for us is how effectively grasses complement flowers in the garden, often looking their best when flowers have the least to offer. Spring is when we are least likely to appreciate grasses, although the fresh green growth of some is a welcome addition to the garden. Some of the best ornamental grasses are the last yard plants to start into growth; *Miscanthus* and switch grass (*Panicum)* do not start until spring has almost turned into summer.

Early summer, the time when so many perennials and roses are looking their best, sees quite a few grasses come into their own, generally the more quietly beautiful ones.

ABOVE RIGHT: The melic (*Melica macra*) is the showiest of this delicate-looking genus, but looks good for only a short time.

RIGHT: *Miscanthus sinensis* 'Malepartus' is one of the best of silver grass (*M. sinensis*) cultivars; this one looks particularly fine in autumn.

ABOVE FAR RIGHT: Even in spring, the dried leaves of silver grass (*Miscanthus*), remaining from the previous year, can be attractive.

FAR RIGHT: A *Miscanthus* with *Eupatorium maculatum* 'Atropurpureum' in the background: a classic late summer combination for a larger border.

The yellow-flushed leaves of Bowles's golden grass (*Milium effusum* 'Aureum') complement purple ornamental garlic flowers perfectly, or in shadier spots the blue flowers of comfrey (*Symphytum caucasicum*). The shade-tolerant melic grasses (*Melica*) also look their best now, delicate flowers nodding above broad fresh green leaves. On the whole, though, the role of grasses at this time is to complement the flowers, those with colored foliage, such as silver-leaved wild rye (*Elymus*) and oat grass (*Helictotrichon*) the most useful alongside the numerous pastel shades of this season.

Late summer and early autumn see grasses play a more important role, but still it is a complementary one. Summer-flowering species like silver grass (*Miscanthus*) and moor grass (*Molinia*) stand among the profusion of perennials that flower now, offering qualities of form and texture alongside the rich flower colors, and perhaps give some respite to the eye. Some of these late flowers are large, like the silver grass (*Miscanthus*), and perform alongside the perennials at their own level. Given that so many late perennials are tall and leggy, it is useful to be able to include lower-growing grasses, like varieties of hairgrass (*Deschampsia*) with their cloudlike seed heads or the oatlike spangle grass (*Chasmanthium latifolium*) at the front of a bed.

ABOVE LEFT: The seed heads of moor grass (*Molinia caerulea* subsp. *arundinacea*) are immensely fine.

ABOVE RIGHT: In late autumn, the delicate seed heads of *Molinia caerulea* 'Fontäne' look stunning.

BELOW: Reed grass (*Calamagrostis* x *acutiflora* 'Karl Foerster'), left, with giant feather grass (*Stipa gigantea*) stand out in a border at the very end of the growing season.

It is in late autumn and winter that grasses really come into their own. While a few collapse at the end of the season, notably several moor grasses (*Molinia*), the majority stand up to the weather remarkably well. By the time the last of the year's flowers have died, the grasses are arguably looking their best. It is not just the wide variety of seed heads but the subtle tones of dead leaves and stems, as soft winter light illuminates a wide variety of tones of fawn and brown. The low, almost sideways, direction of winter

sunlight is particularly effective at making the most of these, as it is at catching the seed heads of pampas grass (*Cortaderia*) and silver grass (*Miscanthus*). Leaving dead stems of grasses and perennials standing provides infinitely more interest in the garden in winter than cutting them all down, as many gardeners have up to now. Birds find not only seeds to eat, but also a host of insect larvae hidden in old seed heads. It is not generally until late winter that the old growth begins to look so messy it needs to be cleared away.

Winter weather such as snow and particularly frost adds a further dimension to grasses. Snow can dramatically emphasize anything left standing, such as the taller grasses or the foliage of bamboos, each leaf supporting a narrow strip of snow. Frost coats a thin layer of ice crystals along every available edge of leaf, stem, and seed head, highlighting details like nothing else does.

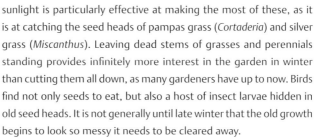

ABOVE LEFT: *Cortaderia selloana* 'Silver Stripe' is one of the lower-growing variegated pampas grasses.

ABOVE RIGHT: Late summer evening sun emphasizes the canes of the fishpole bamboos (from left to right): *Phyllostachys aurea* 'Holochrysa', *P. aurea* and *P.a.* 'Koi'.

THE KEY TO USING GRASSES IN THE GARDEN IS MATCHING THE ENVIRONMENT TO THE PLANT, BOTH IN PHYSICAL AND VISUAL TERMS. HERE WE LOOK AT A VARIETY OF WAYS GRASSES CAN BE USED, SO THE ESSENTIAL CHARACTERISTICS OF EACH ARE DISPLAYED TO THEIR BEST ADVANTAGE AND DIFFERENT ENVIRONMENTS ARE ENHANCED BY THE SPECIES SELECTED.

IN YOUR GARDEN

OPPOSITE: Bamboo is an essential part of any Far Eastern-style garden. This is *Pleioblastus shibuyanus* 'Tsuboi.'

ABOVE: The massive canes of bamboo *Phyllostachys viridiglaucescens* tower skyward.

LEFT: The well-proportioned shape of giant feather grass (*Stipa gigantea*) lends itself to specimen planting, so its elegance can be appreciated. Here it is planted in a container.

BELOW: Many bamboo species can be best appreciated as specimen plants surrounded by an expanse of grass. This is *Fargesia robusta*.

SPECIMEN PLANTING

A specimen plant is one that is grown on its own, in splendid isolation, surrounded by grass or other groundcover, as is the case with a single tree in a lawn. It is the best way to show off a plant whose distinct character is best appreciated when it is grown uncluttered by neighbors. Specimen plants need to look good all over for as much of the year as possible; species that have features that need hiding do not make good specimens. On a smaller scale, the term can also be used to describe plants grown as individuals in containers, which is the most effective way of creating a specimen plant in a small area, rather like putting it on a pedestal. Many of the neat, colored-leaf sedges, such as orange sedge (*Carex testacea*), can look good grown this way.

Pampas grass (*Cortaderia selloana*), is the classic specimen grass, but unfortunately it has been so widely planted in places that are simply too cramped for it, and it has become something of a cliché of suburbia: usually seen planted dead center in the front lawn of a house by someone with delusions of grandeur. The best planting of it I have

LEFT: Chilean *Chusquea culeou* has a dense, clump-forming habit, ideal for specimen planting.

OPPOSITE BELOW LEFT: The linear foliage of a potted *Fargesia murieliae* is contrasted with the naturally rounded forms of pebbles used to line the container.

OPPOSITE BELOW RIGHT: The tangled form of a sedge (*Carex* species) along with Oriental fountain grass (*Pennisetum orientale*) makes an effective container planting combination.

BELOW LEFT: The broad leaves of the bamboo *Indocalamus latifolius* make it a good for specimen planting.

BELOW RIGHT: Snowy woodrush (*Luzula nivea*) is reasonably tolerant of dry shade, making it a potentially very useful plant, especially for the difficult planting conditions found under trees.

ever seen was the use of several clumps together in a loose group in a vast expanse of lawn. Perhaps it is best used as a border plant in most of our gardens, rather than trying and failing to make a good specimen planting of it!

Japanese silver grass (*Miscanthus sinensis*) is an elegant version of pampas grass. Although deciduous, it does make a fine specimen in lawn after a couple of years—it forms a neat clump, and its seed heads form magnificent plumes on tall stems. There are so many varieties in a good spread of sizes it should be possible to find one for any space. The other grass that begs to be grown this way is the giant feather grass (*Stipa gigantea*), whose open bunches of oatlike seed heads hang from long stems flung in all directions; it should be grown in plenty of space to display its characteristic form.

If there is a big enough area of lawn and the site not too exposed or dry, certain bamboos make superb specimen plants. Only those that form tight clumps and do not run are suitable, with those that grow upright and then arch at the top looking the most attractive. Species of *Chusquea*, *Fargesia*, and *Thamnocalamus* are suitable, the latter being particularly useful for soils that do not remain moist through the summer. The black-stemmed *Phyllostachys nigra* can be used as well.

GRASSES IN BORDERS

Borders provide homes for the majority of the plants we grow in our yards. Originally a border was literally that—a strip edging the yard along a backdrop such as a fence, wall, or hedge; the form developed to its climax in Edwardian Britain with grand herbaceous borders, using only perennials, and mostly a selection of late-flowering ones. Since then, the mixed border has become more commonplace, consisting of shrubs and annuals alongside perennials. The 1960s saw the development of island beds, self-contained groups of shrubs and perennials surrounded by lawn or paving that were very different from the traditional border, since there was no backdrop. Some garden designers experimented with other forms of planting that also did without a backdrop, inspired in the U.S. by the prairie landscape, in Britain by the traditional cottage garden, and in Germany by natural wildflower meadows.

The various different forms that borders take have implications for what kind of plants work best in them. The traditional border, where there is a backdrop, and which is appreciated from one side only, works very well for plants that can look scruffy after flowering, or have long bare stems—all of which can be effectively hidden by having other plants grow in front of them to an appropriate height. It also works well for flowers, because the vast majority are highlighted by being seen against a background. But many grasses are arguably not particularly well served by this kind of border; their distinct form is liable to be buried or obscured by surrounding foliage, and there is not enough space to appreciate the patterns created by the radiating flower stems of those like giant feather grass (*Stipa gigantea*) or moor grass (*Molinia*). Upright-growing species with fine seed heads like reed grass (*Calamagrostis*) or many species of feather grass (*Stipa*) are only seen to best advantage against sky. Even in island beds, they can seem hemmed in. It is not surprising that the nature-inspired border styles developed by American and German designers use grasses most extensively and effectively.

BELOW: A border on moist soil in summer: tufted sedge (*Carex elata* 'Aurea') is in the center, with variegated moor grass (*Molinia caerulea* 'Variegata') to the right. Pampas grass (*Cortaderia*) is on the far right.

ABOVE: A variety of grasses—including (from left) hairgrass (*Deschampsia cespitosa* 'Goldtau'), *Achnatherum brachytrichum,* hakone grass (*Hakonechloa macra*) and the leafy green needlegrass (*Stipa calamagrostis*)—transform a border in late summer. The prominent red perennial is *Persicaria amplexicaule* 'Firetail'.

USING DENSE LOW-GROWING GRASSES

Rounded clumplike grasses, especially those with colored foliage, can, however, work well in conventional borders with a backdrop, or in the cottage-garden style, where the juxtaposition of individual plants is very important. They can be very effective used either in blocks containing several individuals of the same variety, where they will make a real, if unsubtle, impact, or dotted throughout a planting to create a sense of rhythm, so the same plant (with its most usefully long season of interest) is repeated at either regular, or random but evenly spaced, intervals. Such repeated dotting of the same plant can create a powerful sense of visual unity in a border, or even in the whole yard, with a strong sense of seasonal continuity as well.

Small dense grasses are particularly useful in confined spaces. A colored-leaf sedge, one of the *Carex* species, can provide variety of color among a small group of other low-growing perennials, while the delicate sprays of flower heads of hairgrass (*Deschampsia*) can complement the broad leaves of a hosta or the rosettes of fern foliage.

Modern low-maintenance plantings can make great use of low colored-leaf grasses when they are used in blocks, alternating with low-growing shrubs like heather and dwarf conifers in exposed situations, or with gray-leaved plants like lavender, sage, and rock rose (*Cistus*), in hot, drought-prone ones. Given their small size, they can be used to create flexible and fluid patterns around the larger plants.

USING SMALL, LIGHT, AIRY GRASSES

These are the grasses that are really likely to get lost, that have to be observed close up to appreciate their sometimes elfin beauty. They should never be planted where they can be obscured by their neighbors, and often look best when there is a whole mass of them together; a patch of any of the melic grasses (*Melica* species) in dappled shade, for example, is a lovely, and arresting sight—whereas one on its own would not even be noticed. Fortunately, most grow easily enough from seed, so it is possible to start from one plant and have many more to plant out within the year.

USING LARGER LIGHT AIRY GRASSES

Big enough to stand alone, but needing space to be properly appreciated, grasses like giant feather grass (*Stipa gigantea*) and moor grass (*Molinia*) look their best in modern open-border style plantings, surrounded by low-growing perennials like stonecrop (*Sedum spectabile*) or *Rudbeckia fulgida* var. *deamii*; dwarf shrubs like thymes (*Thymus*), or other, much lower grasses. Most yards will accommodate only one of these plants; larger areas can be planted with a widely spaced group. They look best with sky behind them, or planted on a slope so their stems hang out into the space above lower plants.

Airy grasses with a less distinctive gowth pattern—more conventionally upright perhaps, like *Stipa capillata* or switch grass (*Panicum virgatum*)—can be combined that much more easily with perennials their own size, but they should not be hemmed in. They can look especially good grown *en masse* and mixed with flowering perennials of a similar light and airy growth habit, such as *Aster lateriflorus*. A small open area can thus conjure up the feeling of a wild meadow.

ABOVE LEFT: Prairie dropseed (*Sporobolus heterolepis*) looks especially wonderful covered in morning dew.

ABOVE RIGHT: Silver grass (*Miscanthus*) combine bulk and majesty with delicacy and elegance. *M. sinensis* 'Africa' is a new variety.

RIGHT: Moor grass (*Molinia caerulea* subsp. *arundinacea*) is a wonderfully delicate tall species for the late summer or autumn border.

USING LARGER UPRIGHT GRASSES

Larger grasses have plenty of presence; whether used singly, widely spaced, or *en masse* is a matter of taste. Reed grass (*Calamagrostis* x *acutiflora* 'Karl Foerster') is stunning in a loose group surrounded by lower perennials or dwarf shrubs like heathers, but can be lost among taller plants. In a dense clump in a modern open yard, they lose some of their individual character but gain something else as a group, a hint of the prairie or a reed bed.

Those grasses, like silver grass (*Miscanthus*), that bear very distinctive flower and seed heads are among the most flexible in use. They create the impression of a reed bed if grown in groups in an open site, but look equally good in a more traditional border setting—their plumes rise above the level of most perennials, and they look good against solid backdrops as well as against the sky. To appreciate them in winter, they need to be in a place where the sun can strike them sideways or backlight them.

USING BAMBOOS

Bamboos have a shrublike bulk, and are evergreen, but have infinitely more elegance than the vast majority of shrubs. Potentially, they are very useful in mixed borders as shrub substitutes, but their need for shelter from wind and hot sun and for a moist soil may limit their use. Some, like *Indocalamus* species, whose main interest lies in their large and distinctive foliage, can be combined easily with a variety of plants, whereas many of the others need to be positioned so their style is not cramped; we need to see the whole canes and the graceful arching of the upper stems. Those that run should only be included if they are enclosed in a bottomless flower pot or sheets of slate.

ABOVE LEFT: Melic grass (*Melica ciliata*) is a relatively small species that adds a touch of the natural landscape to garden plantings.

ABOVE RIGHT: Switch grass (*Panicum virgatum*) is inconspicuous until it flowers.

BELOW: The variegation of the bamboo *Pleioblastus variegatus* echoes an azalea.

RIGHT: Japanese silver grass (*Miscanthus sinensis*) combines with needlegrass (*Stipa calamagrostis*).

FAR RIGHT: Pampas grass (*Cortaderia selloana* 'Silver Fountain') needs space.

USING GRASSES WITH PERENNIALS

Grasses are the natural accompaniment to perennials: after all, most of the sun-loving perennials we grow in our gardens grow naturally as wildflowers in meadows and prairies, habitats that are dominated by grasses. The addition of grasses to a border helps to make the planting look wilder and closer to nature and the soft textures of the grasses are the perfect complement to the shapes and the colors of many flowers.

Garden perennials can be roughly divided into two groups: those that are identical to, or at least close to, their wild ancestors and maintain the proportions, if not the colors, of their forebears, so they are still immediately recognizable as members of the same family; and those that have been intensively hybridized and demonstrate it, having lost the elegant natural proportions of wildflowers. Examples of the latter include many of the delphinium and lupine hybrids, asters, chrysanthemums, and hemerocallis. They look out of place in any situation that is not obviously a garden, and simply do not combine well with grasses. The former, though, almost always look good alongside grasses.

Gardeners have been very selective, perhaps too selective, in choosing wildflowers to grow as garden plants. Recently, however, there has been much experimentation with using a much wider range of wild plants in a border. One of the garden designers who is at the forefront of working with both grasses and wild perennials newly introduced to cultivation is Piet Oudolf in the Netherlands. He is particularly fond of using plants that belong to the umbellifer family (the cow parsley, hogweed, and angelica group)—never colorful in the conventional sense, but which have elegance, delicacy, and sometimes magnificence. Many are, like grasses, major components of natural meadow or woodland-edge habitats. Combined with other traditionally used perennials and grasses, they can be used to create the most wonderfully naturalistic schemes.

USING GRASSES WITH SHRUBS

Shrubs occupy a great deal of space in yards, and when not in flower they can be dull and visually domineering. The late season of grasses makes them potentially good companions, although varieties should be carefully selected. Light and airy species tend not to combine well; those with more bulk such as silver grass (*Miscanthus*) hold their own better visually, while low-growing clump-forming species can be used at the base of the shrub canopy to provide interest at ground level. Shrubs with dark foliage, such as hollies, are a good backdrop for those with light-colored seed heads or pale foliage like the reed grass (*Calamagrostis* x *acutiflora* 'Overdam'). If conditions are right, bamboos mix with shrubs on their own terms in size and scale, their elegant shapes effective next to amorphous shrub shapes.

Given that shrubs can cast shade and they tend to extract considerable quantities of moisture and nutrients from the soil, you should make sure conditions are suitable for grasses before planting around established shrubs. Soil that is full of roots is unlikely to be suitable. Conversely, grasses grow more rapidly than young shrubs, and they should not be planted within thee feet of newly planted stock.

USING GRASSES WITH BULBS

Most bulbs and grasses look their best at opposite times of the year to each other. By careful organization, a succession of interest can be arranged, so bulbs are planted around the base of late-flowering, late-emerging grasses like silver grass (*Miscanthus*) and switch grass (*Panicum*). By the time grasses begin to grow in late spring, crocus and snowdrops are dormant, and most daffodils and narcissi are nearly so. In the autumn, the grasses are at their peak, filling the space that the bulbs take up in the spring.

USING GRASSES WITH ANNUALS

The quickest grasses to establish are those annual species that can be sown in the ground to flower only a few months later, such as the fluffy hare's tail (*Lagurus ovatus*) or the quaking grass, *Briza maxima*. They make excellent companions for pretty cottage-garden-type annuals like love-in-a-mist (*Nigella*), English marigolds (*Calendula officinalis*), or any of the new generation of natural-looking annuals and half-hardy perennials. As with perennials, annuals should be chosen for their wild looks, not top-heavy, highly bred French marigolds (*Tagetes*) or petunias.

EXPOSED AND DIFFICULT CONDITIONS

Grasses are some of the most useful of all plants for those gardens that experience the worst of what the elements can throw at them; wind, salt spray, cold, or drought. The right species do have to be chosen, but it is true to say that grasses are among the most tolerant of all garden plants, and there are some exceptionally good plants for the most testing conditions. Quite apart from their tolerance, the structure of grasses seems almost specifically designed for exposed and other difficult places: there are no boughs or delicate flower stems to break; they bend in the wind to bounce back up again once the tempest subsides; and their growth points are protected by nearly always being at, or near, the base of the plant at ground level, so that if the leaves are broken or eaten, more can be rapidly and plentifully produced.

Cold exposed yards are ideal for a number of grasses from heathland habitats, including true grasses like moor grass (*Molinia caerulea*), and many sedges, including the colored-leaved New Zealand sedgess like *Carex testacea*. The fact that the latter are evergreen is a terrific boon in situations that can be extremely bleak in winter.

Warmer, but still exposed, yards are suitable for practically all grasses, with many withstanding hurricane-force winds. Many grasses also survive in coastal areas that experience winds laden with salt, which can do terrific damage to many foliage plants. Those most suitable are ones with a gray or blue tinge to their foliage, such as the blue oat grass (*Helictotrichon sempervirens*) or species of wild rye grass (*Elymus*); the color indicating the tiny hairs that protect against sun scorch, desiccation, and salt spray. Indeed many species of wild rye grass are originally from maritime environments. The gray tones of the foliage often seem appropriate to coast settings as well, as does the upright character of the foliage, which forms a contrast to flat horizons.

FAR LEFT: The rich brown foliage of clumps of evergreen leather leaf sedge (*Carex buchananii*) is the perfect complement to exposed and weathered stonework in the winter garden.

CENTER LEFT: Sheep's fescue (*Festuca amethystina*) takes on glorious color in the autumn. It is one of many fescues that is ideal for exposed conditions.

CENTER RIGHT: Blue oat grass (*Helictotrichon sempervirens*), like many other blue-gray grasses, survives dry conditions well, which makes it extremely useful.

RIGHT: Wild rye grass (*Leymus arenarius*), an exquisitely blue species, looks especially good in spring, when the new leaves at at their freshest and the color is most intense.

Many grasses are also suitable in areas that can be drought prone, although careful selection of species will have to be made. Again, those with gray foliage are usually good, as well as any plants known to come from areas of steppe, extensive areas of grassy plains, or areas of semi-desert, such as most of the feather grasses (*Stipa*), or from dry prairies, such as the big bluestem (*Andropogon gerardii*). Apart from those species known to be moisture-lovers, most of the grasses are reasonably drought tolerant. Moreover, they usually do not look quite as dreadful as perennials and shrubs can after an extended period of drought, and they often recover faster.

One characteristic of many perennials and shrubs from difficult environments is that they have a spectacular but brief period of flowering; think of the brilliant splashes of color on Mediterranean hillsides in spring, and moorland hills ablaze with purple heather in summer. Grasses, the evergreen ones especially, help to even out this rather unbalanced flowering pattern through the year, providing interest in every season. They also look attractive and appropriate planted alongside the characteristic hummocky shapes of dwarf shrubs such as lavender and heather that tend to be the planting mainstay of such difficult situations.

WATER GARDENS

An essential part of many natural waterside environments is the presence of grasses, such as reed beds fringing lakes and tough clumps of rushes next to mountain streams. And grasses can do much to enhance water features in the yard also, particularly in blurring the edges between the water itself and the rest of the garden. Such a blurring creates the impression that the body of water is natural. Even if it is intentionally not natural, and very formal in design, the soft textures of grasses and their shapes usually complement the clear horizontals of water and edging. Having plants that arch over water is an effect that many find pleasing; many grasses are valuable here, such as the catkinlike flower and seed heads of the weeping sedge (*Carex pendula*), arching out on wiry stems.

In discussing appropriate grasses for yards, the distinction must be made between moisture-loving species and those that need or tolerate ordinary garden soils, but which look like waterside grasses. Moisture-loving species of the true grasses are nearly always vigorous growers with a spreading, often aggressive, habit, making it essential to choose varieties are appropriate to the size of the water feature. Many waterside

OPPOSITE ABOVE: Cotton grass (*Eriophorum angustifolium*) is a plant for acid soils.

OPPOSITE BELOW: Tufted sedge (*Carex elata* 'Aurea') adds a sparkle to the waterside.

ABOVE: SILVER GRASS (*Miscanthus*) is good near water—it evokes wild reeds without their invasive habits.

ABOVE TOP RIGHT: Grasses and water go together naturally—this is made the most of in this garden.

ABOVE RIGHT: A selection of bamboos, some clipped to make a Japanese-style hedge, surround a pond.

BELOW: *Molinia caerulea* subsp. *arundinacea* is a taller, more upright form of the robust purple moor grass.

sedges are competitive runners, too. Given that many familiar waterside grasses are so competitive, it is fortunate that there are smaller versions available of some of them; the distinctive reed maces, for example (*Typha* species), are suitable for only the largest ponds, yet there is a dwarf form—*Typha shuttleworthii*.

Creation of a successful natural-style water garden involves the development of a moisture gradient that runs from the water through the water margin to waterlogged soil, moist soil, and finally dry land. Such a gradient offers a wide range of habitats for plants and for the wildlife that is a feature of a natural pond. Some grasses can be grown in the water itself as marginal plants, such as tufted sedge (*Carex elata* 'Aurea') or species of reed mace (*Typha*). Others are happier in wet ground on the bank, such as ribbon grass (*Phalaris arundinacea*). Then there are those that do best in constantly moist soil, like moor grass (*Molinia*). Many of the latter do as well in ordinary conditions, making them useful for a situation that often occurs: water is enclosed in a liner that does not allow for gradation from water to dry land. Marginals can be sunk into the water's edge, but there is no wet ground. Silver grass (*Miscanthus*), which looks like refined reeds, is ideal for planting next to water, to create the impression of natural marshland.

WILD GARDENS

A key and very obvious distinction between wild plant communities and the garden is the predominance of grasses in so many open habitats, and their almost total absence in the vast majority of yards, except as lawn, kept obsessively mown. It is not surprising that the development of a more natural garden style has gone hand in hand with an increase in the appreciation of grasses, and the linking of natural inspiration to gardens and landscapes by designers such as James van Sweden in the U.S., Piet Oudolf in the Netherlands, and Urs Walser in Germany. Starting with the lawn, why not let it grow a little, to encourage low-growing wildflowers like daisies and cowslips? Many gardeners have taken this a stage further, by trying to establish wildflower meadows as alternatives to lawns. Unfortunately, the majority of attractive meadow wildflowers are natives of thin European grassland and are rapidly swamped by grasses on the majority of soils. Even if slow-growing, non-competitive grasses are used, aggressive species often seed themselves in naturally after a number of years, and on most soils then take over. However, for thin soils, or a substrate of sand or crushed rubble, they, and their accompanying limestone flora grasses, are immensely suitable. Some of the grasses are quite beautiful in themselves, such as the perennial quaking grass (*Briza media*).

On most soils of average fertility and moisture content, where it is difficult to establish meadow wildflowers, simply letting coarse lawn grass grow can result in areas of long grass of great beauty. Such areas can be made to look intentional, rather than just neglected, if paths of mown grass are made through them, or cut around them, or their context is a relatively formal one.

Regions of clearly defined continental climates are most suitable for prairie-style plantings; clump-forming grasses such as big bluestem (*Andropogon*) are used as a matrix for tall late-flowering perennials. Natural prairie has a highly fertile soil and a flora of incredible richness and diversity. Prairie restoration is becoming popular as a garden and landscape style in the Midwestern states: it is good for wildlife and needs less water than conventional planting.

Borders can be made to look more natural by using grasses, almost any grasses, that contribute an air of natural insouciance, especially those with prominent flowers or seed heads. The use of multiples of one or two grasses scattered throughout a planting is particularly effective at hinting at meadow or prairie, with their natural close association of perennials and grasses.

The most effective grasses at developing a natural feel are those that look as if they could be superior natives of the area, rather than the exotic looking, or obviously unnatural, like variegated varieties. Species of reed grass (*Calamagrostis*), fountain grass (*Pennisetum*), and feather grass (*Stipa*) are highly attractive and they fit easily into a wide variety of landscapes.

LEFT: Melic grass (*Melica transsilvanica*) grows alongside magenta cranesbill (*Geranium sanguineum*) and other drought-tolerant wild plants.

ABOVE: The extraordinarily long awns of feather grass (*Stipa pulcherrima*) wave in the breeze among a selection of steppe, or dry grassland, species.

OPPOSITE: Wild grasses evoke a sense of romance, though some regard them as messy.

BELOW LEFT: Variegated velvet grass (*Holcus mollis* 'Albovariegatus') looks especially bright in spring.

BELOW RIGHT: Clumps of tufted sedge (*Carex elata* 'Aurea') are used to edge a path.

BOTTOM LEFT: Silver hairgrass (*Koeleria*), though not a runner, makes effective groundcover for shade.

BOTTOM RIGHT: The perennial quaking grass (*Briza media*) can be used in meadows on drier soils.

FAR RIGHT: Greater woodrush (*Luzula sylvatica*) is excellent evergreen groundcover for difficult shade.

GROUNDCOVER

Lawns are the obvious and most common use of plants as groundcover; planting which is functional as much as it is decorative, aiming to please as an overall impression rather than with detail. Alternatives to grass as groundcover have traditionally been used only in areas where lawn grass will not grow—chiefly the use of reliable evergreens such as ivy in shade. The varieties chosen have tended to be dark green, dull, and to have given the concept of groundcover rather a bad name in some quarters. Recently, more adventurous use of groundcover plants, especially in the U.S., has encouraged a more positive outlook. So what is grass as a groundcover if it is not lawn?

Lawn is maintained at an artificially low height by mowing, whereas groundcover grasses are allowed to grow to their normal height, which is usually fairly low. Groundcover in shade tends to be functional, but it can be decorative: woodrushes (*Luzula* species) are spreading evergreens tolerant of dry shade, and they are moderately attractive. The more lush and attractive palm sedge (*Carex muskingumensis*) is one of several suitable species of sedges that thrive in damp shade. The most elegant groundcovers for light shade is melic grass (*Melica*), which can form a lawnlike growth to a height of about a foot when flowering. Their impression is one of great delicacy, but melic does not spread like lawn grasses or many sedges. Large numbers of plants would need to be raised from seed to create much of a spectacle.

Groundcover in sun tends to be chosen on more positive grounds than that in the shade. Low-growing grasses like varieties of blue fescue (*Festuca glauca*) have been the most commonly used until now, but there is no reason the taller species should not be used, perhaps putting the running habit of Lyme grass (*Leymus arenarius*) to good use. Species of fountain grass (*Pennisetum*) are used extensively as groundcover, their fluffy flower/seed heads flopping and meshing together by late summer to make up for the fact their clumps of leaves do not. However, they are only reliable as groundcover for areas with a continental climate.

As an alternative to the grayish tones predominating among many of the sun-loving grasses that are suitable for use as groundcover, the warm browns of the New Zealand sedges should be considered. Low-growing and brown *Carex comans* is the most suitable, but others are worth considering. When planning groundcover, it is worth considering whether you want a uniform look, with solid blocks of multiples of the same variety of grass, or a more naturalistic blending of different varieties. Both can be highly effective ways of contrasting different foliage colors, forms, and textures, the former looking more appropriate in urban surroundings, the latter in rural ones. The blended approach only really works if plants of roughly similar height are chosen, whereas if block planting is done, contrasts in height between different blocks can be part of the overall effect.

ABOVE: Blue fescue (*Festuca glauca*) makes a strong contrast to the purple foliage of the Japanese maple (*Acer palmatum* 'Atropurpureum').

RIGHT: *Shibataea kumasasa* is a very effective dwarf bushy bamboo for small spaces.

CENTER RIGHT: The bolder gardener should consider how effective large grasses such as Japanese silver grass (*Miscanthus sinensis* 'Ghana') can be in small gardens.

FAR RIGHT: Bamboos are surprisingly effective in confined as well as large areas, as this marble bamboo (*Chimonobambusa marmorea*) and *Thamnocalamus tessellatus* demonstrate.

SMALL SPACES

Small spaces in yards are difficult because every plant tends to be on view the whole time; there is no room for anything that looks messy for very long. Plants with a long season, such as grasses, are inherently more suitable than short-lived wonders, however spectacular. Confined areas can also be difficult because the prevailing conditions are often somewhat unfriendly to many plants: shade or hot sun, depending on the aspect of walls; drafts and gusts of wind tearing around corners, rubble-filled beds at the foot of foundations. Grasses, with the exception of bamboo, are well suited to many trying conditions, which makes them even more suitable for such places.

Views on planting small places differ. I share the opinion of those like designer James van Sweden (*see page* 37), that small places need dignifying and magnifying through the use of large plants; big silver grass (*Miscanthus*) and bamboo. But there are those, especially those who love collecting plants for their own sake, who favor the use of a larger variety of small ones. It all comes down to personal taste.

Traditional Japanese courtyards often feature bamboo, and where there is shade and a reasonably moist soil, or the certainty of being able to keep them watered, they make some of the finest plants for confined places. The height of many varieties can create the impression of being in a forest. Others are useful for the fact that, unlike woody shrubs, they are inclined to be tall and thin and thus fit into tight spaces better. There are also attractively bushy miniature species like *Shibataea kumasasa*.

Grasses with details that go unnoticed in the wider yard are suitable for places where there is little else to distract attention: the large seed heads of spangle grass (*Chasmanthium latifolium*); the delicacy of melic grasses (*Melica*), or the variegation of Japanese sedge (*Carex morrowii* 'Variegata'). Positioning them near seating is always a good idea. Grasses that start into growth and flower late are not so good where space is limited. More valuable are compact species that are evergreen or that grow early in the season. The low but elegant hakone grass (*Hakonechloa macra*) has become a favorite with many garden designers for this reason.

Whatever their size, plants chosen for small spaces do need to be those that do not spread rapidly, or if they are, they must be contained. Pots from which the bottoms have been removed or tubelike building materials, such as terracotta flue liners, are often a practicable and attractive way of growing a plant while limiting its sideways spread. Less attractive materials can be used below ground.

SHADE

Grasses, as we have seen, are amazingly tolerant of a very wide range of conditions, more so than almost any other groups of cultivated plants. But reduced light is their Achilles' heel. Their highly efficient photosynthesis mechanism, which is one reason for their global dominance, is rendered largely ineffective out of direct sunlight. Walk from any area of grass, meadow, or lawn under trees, and you will notice how rapidly the grass thins out. In full shade it dies out completely.

There are just a few true grasses that are tolerant of light shade. Bowles' golden grass (*Milium effusum*) flourishes, as do the various graceful species of melic (*Melica*), which we have already met in Groundcover (*see page* 65). But the relatively delicate melic need to be grown *en masse* to be effective and to survive the competition of more aggressively spreading shade-tolerant plants, so they are really useful only as ground-cover, not as anything more spectacular. Spangle grass (*Chasmanthium latifolium*) tolerates some shade and is a good contrast to more graceful species.

Looking beyond true grasses, though, we find a good number of sedges that thrive in the shade of trees and buildings: the dark green palm sedge (*Carex muskingumens*), for example, or the solid clumps of weeping sedge (*C. pendula*), or the lower growing plantain-leaved sedge (*C. plantaginea*). These are all evergreen and can be combined with other shade-loving plants to provide a rich tapestry of different leaf shapes; one of the joys of gardening in shade is the greater variety of leaf shapes and textures available. The fact that they are evergreen is very useful for maintaining interest through the year, especially since shade plants tend to be spring- or early summer-flow-ering, with little happening later. Variegated ones like variegated Japanese sedge (*C. morrowii* 'Variegata') are useful for bringing in some light at all times.

The woodrushes (species of *Luzula*) are also evergreen and shade tolerant, doing far better in dry shade than many other plants. They are useful to form bulk planting, but mixed in with smaller numbers of other, more immediately attractive plants, perhaps hellebores, bulbs, digitalis, or in more moist shade, ferns and hostas.

Bamboos are potentially very useful plants for light shade, but only if the soil never dries out. Shrubs often do not flourish in shade, but bamboo offers a good alternative if height and bulk are needed. Those with running roots, like species of *Sasa*, can be put to good use if there are large areas that need to be filled. Those that do not run, but form tighter clumps, are perfect for shady situations that need filling upward, but not outward, as often happens around buildings, walls, and next to paths. The wide variety of foliage, shape, and texture among bamboos makes them exceptionally rewarding plants to bring a sense of style to shade.

BELOW FAR LEFT: Weeping sedge (*Carex pendula*) is an unusual plant that is good for difficult dry shade.

BELOW CENTER LEFT: Snowy woodrush (*Luzula nivea*) forms a carpet of narrow evergreen foliage in shade.

BELOW LEFT: *Luzula sylvatica* is the most robust of the woodrushes for use as groundcover.

BELOW: *Deschampsia cespitosa* 'Goldschleier' is a cultivar of a hairgrass that does well in dry light shade, and, if densely planted, it makes effective groundcover.

GRASSES IN CONTAINERS

The use of ornamental grasses in containers is a relatively novel approach, but one that has immense potential. Flowering plants in containers tend to be stuffed in cheek to cheek, a situation that can often lead to a sense of overcrowding, hence the long-standing use of strongly colored foliage plants to complement them. Grasses can play a similar complementary role, either through their foliage color, or the form of the flower/seed head or, indeed, the form of the whole plant.

The fact that many grasses have an arching habit makes them eminently suitable for container culture anyway: the shape often looks just right in a pot, with no other plants needing to be being involved, unlike stiffly upright or trailing plants, which always need at least one other shape with them to complement them. Those with fine leaves that can infiltrate and mix together with the stems and flower heads of other pot denizens are the most useful. Those that form squat clumps or have stiff or broad leaves are

much less so. Their long period of interest is a great help at the beginning and the end of the season, before and after flowering plants are at their best, and to provide continuity through the waxing and waning of the flowering species.

Fresh young growth or evergreen leaves in spring make a good complement to flowering bulbs and early perennials or bedding plants, such as primroses and daisies (*Bellis perennis*). The yellowish-green young leaves of the tufted sedge (*Carex elata* 'Aurea') or Bowles's golden grass (*Milium effusum* 'Aureum') combine well with the intense blues of grape hyacinths. Blues are also effective next to dark bronze colors, such as that of the hook sedge (*Uncinia rubra*).

Summer containers can be enlivened with a huge variety of grasses. Annual grasses can be grown from seed or divisions, or young plants of perennial species can be used, although it is vital that the latter are not strongly spreading; otherwise, you will end up with a pot of grass and nothing else. Certain annuals are notable for their highly

FAR LEFT: If they are kept well-watered, bamboos make very good container plants. This is *Pseudosasa japonica* 'Tsutsumiana'.

CENTER LEFT: Hakone grass (*Hakenochloa macra* 'Aureola'), on the far left, works well as an accompaniment to gravel and plants in pots.

BELOW LEFT: Japanese variegated sedge (*Carex morrowii* 'Variegata') is a cultivar whose neat shape lends itself to container cultivation.

LEFT: Container growing is recommended for those who worry about the invasive habits of many bamboos, such as *Sasaella masamuneana* f. *albostriata*.

large seed heads, such as quaking grass (*Briza maxima*), or conversely for fine seed heads, which form clouds among the more clearly defined shapes of flowers; examples include little quaking grass (*Briza minor*) and cloud bent grass (*Agrostis nebulosa*).

The pastel color combinations favored by many can be given an extra dimension with the striking blue leaves of wild rye grass (*Elymus magellanicus*), or the pale fawn, fluffy seed heads of fountain grass (*Pennisetum alopecuroides*). Warmer combinations can include any of the bronze or yellow sedges such as *Carex dipsacea*, or the golden horizontal bands of porcupine grass (*Miscanthus sinensis* 'Strictus'). Evergreen grasses are particularly important in containers. Japanese sedge (*Carex morrowii* 'Evergold') is good not just because of its color, but because its expansive shape makes up for what squat or ground-hugging winter flowers, such as pansies or heathers, lack.

ORIENTAL STYLE

Bamboo has a particularly strong association with the Far East, although it is far more widely distributed than this. Their role in the gardens of China, Japan, and Korea, and in the arts of these countries, accounts for this association, and it is thus rather obvious a point that using bamboo seems to be a vital element in creating Oriental-style gardens. However, far too many imitations have been made that seem to bear little resemblance to the nobility of the originals, and depend too much on the almost knee-jerk reaction so many Westerners have: bamboo equals Japan and China. The best Oriental gardens are those that contain none of the obvious plants, but local species chosen for the way they echo the spirit of traditional Oriental plants. An alternative is to be inspired by the style of Oriental gardens, but to develop gardens that do not pretend to be like them, but that share certain characteristics, such as simplicity or elegance. Such has been the approach of those few contemporary designers who have made successful gardens for modern architectural surroundings.

Grasses can play a vital role in both approaches. Those who have really understood what Japanese gardens are all about, rather than the superficial form they take, will select plants on the basis of elegance and economy of form. Grasses, bereft of brightly colorful flowers, are eminently suitable for such gardens. Japanese silver grass (*Miscanthus sinensis*) is a native Far Eastern plant, and yet it is far less readily associated with China and Japan than is bamboo, making it ideal for more subtle suggestion. Its graceful one-sided flower/seed panicles are invaluable in autumn and winter. Weeping sedge (*Carex pendula*), with its catkinlike heads pendent on wiry stems, is a very good example of the kind of plant that makes a statement based on understatement, as are melic grasses (*Melica*) too, which have the advantage that they grow well in shade.

Courtyards, where few plants are used, are the perfect place to draw attention to subtle details. A clump of grass that might go unnoticed elsewhere could become the center of attention or play a role in counterpointing something else. In such situations, interplay between plant form and texture, and that of surroundings—fences, walls, and seats—becomes the more vital.

OPPOSITE: Hakone grass (*Hakonechloa macra* 'Aureola') grows alongside bronze sedges in a Japanese-style garden.

ABOVE LEFT: Bulbous oat grass (*Arrhenatherum elatius* subsp. *bulbosum* 'Variegatum') is particularly effective in spring.

ABOVE CENTER: Bowles' golden grass also looks best in spring.

ABOVE RIGHT: Japanese variegated sedge tolerates light shade.

RIGHT: Trimmed bamboo kept in shape makes a traditional-style edging.

ABOVE: Umbrella sedge (*Cyperus*) grows against a background of clipped bamboo in mid-winter.

BELOW: The young growth of velvet grass (*Holcus mollis* 'Albovariegatus') spreads among pebbles and gravel.

BELOW RIGHT: Cultivars of Japanese silver grass (*Miscanthus sinensis*) are an essential part of modern style.

BELOW FAR RIGHT: *Pennisetum alopecuroides* 'Herbstzauber' is one of many highly effective fountain grasses, which tend to do best in mid-continental climates.

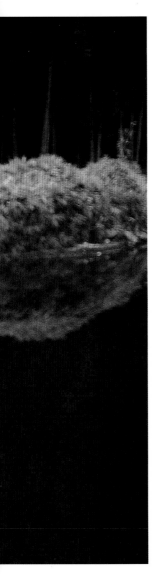

CONTEMPORARY STYLE

One of the striking things about grasses is how recent is their entry into the garden. I have come across a reference to using ornamental grasses in an 1880s magazine, but the writer was clearly a voice in the wilderness, and it was not until the 1950s that gardeners in the U.S. and Germany started using the perennial species, with real popularity not coming until much later. They are thus inescapably contemporary plants.

One of the things I love about grasses is the way they fit so well into contemporary environments. They have a simplicity and purity of form that blends well with modern architecture or abstract sculpture, softening the hard outlines of glass and steel buildings, and introducing a note of nature into urban places. It is no coincidence that bamboos and certain grasses, notably Japanese silver grass (*Miscanthus sinensis*), were portrayed by Japanese artists, and it is Japanese gardens that have inspired much of the best contemporary garden and landscape design. Traditional Japanese art captured the way these plants move so well, and it is their almost constant motion in the slightest breath of wind that makes them so valuable in the vicinity of buildings. The absence of a wide range of color in Japanese gardens makes us concentrate on the form of plants, and much the same is true of modern-style gardens; a few especially elegant plants are chosen and combined with natural elements like pebbles or weathered wood, or manufactured ones such as paving.

It is a special quality of grasses that they are equally effective used *en masse* or as individuals. Those with strong form, like silver grass (*Miscanthus*) or feather grass (*Stipa*), are as highly effective in planters in a tiny roof terrace as in great masses along a driveway or around a building. The effects created will be different, but both will be equally striking. It was the Brazilian landscape architect Roberto Burle Marx who developed the most truly modernist garden style, using plants in ways designed to emphasize the contrasts they had with each other as strikingly as possible. Oehme and van Sweden in the U.S. have very much followed in his footsteps, using large quantities of grasses like reed grass (*Calamagrostis*) or fountain grass (*Pennisetum*) to contrast with drifts of flowering perennials or shrubs. But they also know when to use individuals or small quantities to equally great effect, for example using three clumps of moor grass (*Molinia*) in a drift of stonecrop (*Sedum spectabile*), the arching stems making an airy contrast with the dumpy stonecrop. It is this playing off of a limited number of different plants with each other that contributes such dynamic creative energy to the best contemporary gardens.

RIGHT: Reed grass (*Calamagrostis* x *acutiflora* 'Karl Foerster') makes an effective upright statement over a long season.

BELOW: A variety of grasses combine effectively with perennials in late summer in the garden of Dutch designer Piet Oudolf.

BOTTOM: *Miscanthus sinensis* 'Malepartus', a cultivar of Japanese silver grass, rises above a carpet of palm sedge (*Carex muskingumensis*) in a bracelet of brick and yew, a highly effective simple modern design by Piet Oudolf.

OPPOSITE: The bromes are all attractive grasses, somewhat underrated commercially—this is *Bromus rigidus*. They tend to be short-lived, but are easily propagated by seed.

RIGHT: The canes of the bamboo *Phyllostachys edulis* var. *heterocycla* 'Kikko' are very strangely shaped. It is hardy only in areas with a mild winter.

HERE WE LOOK IN DETAIL AT THE BEST OF THE MORE WIDELY AND EASILY AVAILABLE GRASSES, BAMBOOS, SEDGES, AND WOODRUSHES. THIS DIRECTORY LISTS OLD FAVORITES AS WELL AS TRYING TO GIVE A FLAVOR OF THE FUTURE—THIS IS A VERY DYNAMIC AREA IN GARDENING, AND MORE GOOD VARIETIES APPEAR EVERY YEAR.

DIRECTORY

All non-bamboos are deciduous unless otherwise indicated. Bamboos (B) are all evergreen. Figures for Height (H) and spread (S) indicate a plant's maximum size when fully grown.

1

2

ACHNATHERUM BRACHYTRICHA
(SYN. CALAMAGROSTIS BRACHYTRICHUM OR STIPA BRACHYTRICHUM)

Pale brown flower/seed heads with a mauve tinge in late summer and autumn over clumps of fresh green foliage. Zones 6–9 H 30 inches x S 20 inches.
CULTIVATION: Most fertile soils in full sun.
USES: Effective with *Monarda* (bee balm) in mid- to late summer or purple and yellow perennials later.
SIMILAR VARIETIES: *A. calamagrostis* has more romantically drooping, yellower seed heads.

ALOPECURUS PRATENSIS 'AUREOVARIEGATUS'
Foxtail grass

Orange-yellow and green striped foliage forms mounds. Pokerlike flowers are tinted purple in early summer. Zones 6–9 H 8 inches x S 16 inches.
CULTIVATION: Full sun or very light shade in fertile soil.
USES: Edging or low carpets. Shallow containers. Plant with fescue (*Festuca*) or black mondo grass (*Ophiopogon planiscapus* 'Nigrescens.')

ARUNDO DONAX
Giant reed

Huge gray-green leaves often 30 inches long. Large autumn plumes of red-tinted panicles on mature plants if left unpruned, in warmer climates only. Zones 6–10 H 10–15 feet x S 3–10 feet.
CULTIVATION: Any fertile and moist, but well-drained soil in full sun. Tolerant of wind and heat once established. Will withstand frost to 14°.
USES: Windbreak for coastal gardens. Feature plant for combining with other large-leaved plants.

BASHANIA FARGESII (B)

Distinct gray glaucous canes, eventually thick. Medium pointed leaves on short, rigid horizontal branches. A very drought-tolerant bamboo with a deeper root structure than most. Establishes quickly even in dry, stony soil.

3

Zones 5–9 H 10–12 feet x S 10–15 feet.
CULTIVATION: Full sun, but will tolerate semi-shade. Spreads quickly with plenty of moisture.
USES: Screening, large grove, or windbreak.

BRIZA MEDIA
Perennial quaking grass

Short and clumping with many heart-shaped golden flowerets held above fine green leaves. Zones 4–10 H 12 inches x S24 inches
CULTIVATION: Form dense colonies through slow spread and self seeding. Dislikes competition. Easy in fertile soil in sun or light shade. Does best on dry limestone soils.
USES: Good wild garden grass or underplanting for taller grasses, perennials, or shrubs. Combine with hairgrass (*Deschampsia*) or cranesbill (*Geranium*).

BROMUS INERMIS 'SKINNER'S GOLD'

Brome

Very bright yellow-green foliage and tall drooping yellow flowers on pale creamy stalks in midsummer.
Zones 4–8 H 3 feet x S 3 feet.
Cultivation: Poor soils will keep the plant short, refined and colorful. Full sun.
Uses: Wild areas, mixed borders or dense grass plantings. Use as a foreground to reed grass (*Calamagrostis*) or silver grass (*Miscanthus*), so it will brighten the early foliage of these late-flowering plants.

CALAMAGROSTIS X ACUTIFLORA 'KARL FOERSTER'

Reed grass

Dark narrow foliage is topped by stiff, upright bronze-purple flower spikes, which age to buff-gray. Stands winter and winds well.
Zones 5–9 H 6 feet x S 3 feet.
Cultivation: Moisture-retentive soil in sun.
Uses: Background, screen, or focal point. Group with lower-growing perennials, e.g. day lilies (*Hemerocallis*) or cranebill (*Geranium*), or with other looser grasses.
Similar varieties: *C.* x *acutiflora* 'Stricta' is almost identical to 'Karl Foerster.' *C.* x *acutiflora* 'Overdam' is variegated.

CAREX

Sedge

Large genus with 2,000 species from a wide range of habitats. Species vary in shape and size from short, dense tussock-forming plants suitable for large rock gardens or screes to more robust wilderness plants for bog gardens or meadows.

CAREX ATRATA

Stiff, blue-gray evergreen leaves forming dense carpets. Dark purple flowers produce jet-black seed heads that will last throughout the summer.
Zones 4–9 H 18 inches x S 24 inches.

Cultivation: Full sun or open aspect in light shade, average to moist soil.
Uses: Gravel-mulched areas, containers, or low carpeting. The short leaves look effective at a pond edge with irises.
Similar varieties: *C. caryophyllea* 'The Beatles' is similar in habit, but forms a tighter clump.

CAREX COMANS

New Zealand sedge

Arching, brown hairlike foliage forms evergreen tufts.
Zones 7–9 H 12 inches x S 24 inches.
Cultivation: Full sun in average soil.
Uses: Very versatile. Use in containers, mixed planting or *en masse* with other grasses. Good with bronze, blue, and purple.
Similar varieties: The bronze form of *C. comans* is shorter, with dusky chocolate-brown foliage. *C* 'Frosted Curls' is taller with unique blond coloring. *C. flagellifera* has deep chestnut-brown hairlike foliage.

[1] *Achnatherum brachytricha* is a first-rate medium-sized grass.

[2] *Arundo donax* is a hardy giant reed from the tropics.

[3] *Bashania fargesii* has characteristic leaf pleating.

[4] *Briza media* is a species of thin limestone grassland.

[5] Reed grass (*Calamagrostis* x *acutiflora* 'Karl Foerster') has a long season.

[6] The rich brown foliage of New Zealand sedge (*Carex comans*) is much appreciated in winter.

CAREX ELATA 'AUREA'
Tufted sedge

Rich golden-edged, long narrow leaves for early spring and summer color on tight, slowly spreading clumps. Dark brown flowers in early summer.
Zones 5–10 H 28 inches x S 16 inches.
Cultivation: Sun or light shade, moist soil or pond margins.
Uses: Best admired as a solitary specimen.

CAREX HACHIJOENSIS

Glossy evergreen leaves, slightly arching. Pale flowers.
Zones 6–9 H 16 inches x S 16 inches.
Cultivation: Most soils, sun or shade.
Uses: This species and similar variegated forms (below) are good low cover in between larger, clump-forming grasses or other low perennial planting.

CAREX MORROWII 'VARIEGATA'
Variegated Japanese sedge

Pale, subtle variegation creates an all-over milky green color. Rigid evergreen leaves, slightly arching.
Zones 5–9 H 14 inches x S 20 inches.
Cultivation: Full sun or shade, moist or dry soil.
Uses: Will cover ground under dark evergreens with ample watering to establish young plants. Useful basal planting for bamboos or as formal path edge planting.

CAREX MUSKINGUMENSIS
Palm sedge

Exotic-looking foliage with tall stems and thin leaves arching horizontally with a palmlike effect. Deciduous.
Zones 4–9 H 32 inches x S 20 inches.
Cultivation: Moist shady areas or in a bog garden, although it will tolerate average soil conditions.
Uses: Focal point or groundcover in wild or marginal grass plantings. Bold colors contrast well.

CAREX PENDULA
Weeping sedge

One of the largest sedges, with broad pleated, semi-evergreen leaves, long and arching. Red-brown catkin-like flowers in spring hang from tall arching stems.
Zones 5–9 H 3 feet x S 3 feet.
Cultivation: Moist and rich soil in dappled shade, but likes full sun if kept moist. Can self-seed too generously.
Uses: Woodland path edging or waterside wild gardens; interplant large drifts with foxgloves or columbines.

CAREX PLANTAGINEA
Plantain-leaved sedge

Tightly clumping evergreen with broad, pleated leaves up to an inch wide, with striking spikes of rusty red and black flowers.
Zones 4–9 H 12 inches x S 16 inches.
Cultivation: Best suited to shady woodland, if not too dry. Also rewarding in pots, given ample food and water.
Uses: Superb edging to woodland paths or shady pools.
Similar varieties: Though deciduous, *C. siderosticha* 'Variegata' is similar to *C. plantaginea*, with bold white striped and margined leaves.

CAREX RIPARIA 'VARIEGATA'
Greater pond sedge

Early spring growth is almost completely white, toning to a milky green as the season advances. The early, almost jet-black flowers are a sharp contrast. Striking, desirable, and easy, but very invasive.
Zones 5–9 H 20 inches x S 40 inches.
Cultivation: Sun or light shade is necessary for the brilliant color. Tolerant of all except the driest soils and well suited to bog conditions.
Uses: As a bold and spreading feature on large pond margins where it will bind the soil. Usually too vigorous to combine with other plants, unless they are equally robust, e.g. giant rhubarb (*Gunnera*) or *Ligularia*.

CAREX TESTACEA

Tight evergreen clumps of olive-green hairlike foliage. The color turns a darker orange-red in full sun in winter.
Zones 6–9 H 16 inches x S 24 inches.
Cultivation: Resents hard pruning when mature. Full sun in moist, free-draining soil.
Uses: Tumbles over raised beds or rocks. Compact source of color in winter, or with yellow or blue bulbs in spring.
Similar varieties: *C. dipsacea* is very similar but is more tolerant of wetter conditions.

CHASMANTHIUM LATIFOLIUM
Spangle grass, sea oats

Slowly spreading clumps of broad deciduous leaves. Flat spikes of green flowers turn copper and then gray as the foliage simultaneously changes to pink and straw yellow.
Zones 5–9 H 36 inches x S 24 inches.
Cultivation: Good soil in light shade is ideal. Will grow in full sun but requires more moisture.
Uses: Fine winter garden plant—looks superb in frost.

CHIMONOBAMBUSA MARMOREA (B)

Marble bamboo

Loose clumps of pale green canes turn plum purple, with small twisted leaves in dense bunches at the ends. New shoots are strikingly marbled.
Zones 7–10 H 6 feet x S 10 feet.
Cultivation: Naturally grove forming, but is relatively easy to control because of shallow rooting. Needs good soil, well mulched, in sun or part shade.
Uses: A versatile plant for waterside or wild places. Fine and easily manageable in containers.
Similar varieties: *C. tumidissinoda* is elegant but invasive, ideal for pots, growing to 10 feet in a bed, 3 feet in pots.

CHIONOCHLOA RUBRA

Tussock grass

Narrow ocher-red, evergreen leaves forming dense arching clumps with good red and pink winter tints.
Zones 7–9 H 30 inches x S 12 inches.
Cultivation: Slow to establish. Likes sun in good, well-drained soil.
Uses: The essential winter garden plant. Use with dwarf pines and prostrate blue junipers. Some beach pebbles and driftwood will add background gray tones to a orange, blue, and green.

CHONDROSUM GRACILE

(SYN. BOUTEOUA GRACILE)
Mosquito grass

Unusual flower heads grow at right angles to the vertical stems, silvery at first and then turning purple.
Zones 5–10 H 16 inches x S 16 inches.
Cultivation: Well-drained soil in an open aspect.
Uses: Best used as underplanting around established shrubs that will take up any moisture quickly, providing the ideal soil conditions for this grass.

CHUSQUEA CULEOU (B)

A bold and desirable clump-forming species with thick, yellow-green solid canes, often purple-black when young. Dense, rigid clusters of short, dark evergreen leaves create bottle-brush formations on the upper half of the canes.
Zones 6–9 H 20 feet x S 10 feet.
Cultivation: Good drainage and an open aspect in sun.
Uses: One of the boldest features for a medium to large yard and deserves a prominent placing. Underplanting will spoil the effect of the new emerging canes.

[1] Elegant Japanese variegated sedge (*Carex morrowii* 'Variegata') is a widely available.

[2] A unique pattern of growth makes the palm sedge (*Carex muskingumensis*) a valuable and distinctive ornamental plant.

[3] Weeping sedge (*Carex pendula*) is a tough survivor for difficult areas, especially shade.

[4] Greater pond sedge (*Carex riparia*) is normally vigorous, but variegation (this is 'Variegata') reduces its growth rate.

[5] The bamboo *Chimonobambusa tumidissinoda* is traditionally used for making canes and walking sticks.

[6] *Chusquea culeou* is the best tight clump-forming bamboo.

CORTADERIA SELLOANA

Pampas grass

A variable species. Coarse evergreen, arching foliage with huge silky white plumes standing well above the foliage during autumn.
Zones 7–10 H 10 feet x S 3 feet.

Cultivation: All forms require a sunny aspect in well-drained soil, not too poor. Hard pruning in spring helps to keep the plant neat. Very tolerant of salt and exposure to cold winds.

Uses: Large groups are very effective. Helps to brighten up the foreground of large, formal evergreen hedges and the monotony of huge areas of lawn. Drifts of elephants' ears (*Bergenia*) are a good complement or use it as spot planting with yucca or purple spurge (*Euphorbia*).

Cultivars: *Cortaderia selloana* 'Aureolineata' (syn. *C. s* 'Gold Band') has excellent gold-margined foliage on upright clumps. *C. s.* 'Sunningdale Silver' has enormous silvery-white plumes held well above the foliage.

CYPERUS ERAGROSTIS

Umbrella sedge

Strong clumps of rich green foliage, with deeply veined leaves and large umbels of yellow-green flowers in summer and autumn.
Zones 8–10 H 2 feet x S 2 feet.

Cultivation: Withstands sun or shade, boggy or ordinary soil. Annual clear-up and thinning of self-sown seedlings may be necessary.

Uses: Pond or lake margins in large groups with other marginal plants. Let it seed itself in the larger gravel garden or plant in containers near small formal pools or on the patio.

DESCHAMPSIA CESPITOSA

Hairgrass

The densely tufted evergreen foliage is neat and fresh, and the early maturing golden or bronze flowers last for the greater part of the year.
Zones 4–9 H 36 inches x S 16 inches.

Cultivation: Light shade with fertile, moist soil is ideal. Divide occasionally and shear annually in winter.

Uses: Plant in large drifts or use for spot planting. Extremely versatile, looking good almost anywhere. Combine with hostas, Solomon's seal (*Polygonatum*), and broader-leaved grasses.

Cultivars: *Deschampsia cespitosa* 'Bronzeschleier' has dark, rusty-brown arching flowers and is taller than most cultivars. *D. c.* 'Goldtau' is shorter, with stiff spikes of pale golden flowers lasting for ages and compact dark green foliage. *D.c.* 'Goldshleier' is fairly tall, with golden flower heads.

DESCHAMPSIA FLEXUOSA

Wavy hairgrass

Very fine-leaved and tufted species thriving best in cool climates where the evergreen, shiny needlelike foliage stays lush. The flowers erupt from the foliage up to 14 inches and turn rich purple-brown.
Zones 4–9 H 14 inches x S 16 inches.

Cultivation: Easy and adaptable in all except the most arid conditions, and good on acid soils. Can self seed.

Uses: Good companion to heathers and other low-growing plants tolerant of acid soils.

Cultivars: *D .f.* 'Tatra Gold' is a shorter golden-green form, very bright in spring.

ELYMUS HISPIDUS

Wild rye, lyme grass

Erect leaves of a stunning silver-blue on young growth, arching later in the season. Wheatlike flowers. Clump forming and not invasive, like many others in this genus.
Zones 4–9 H 40 inches x S 16 inches.

Cultivation: A dry stony site with full sun will really bring out the vivid color.

Uses: Dry gravel gardens or sandy areas. Good against a dark background or plants such as purple smoke tree (*Cotinus*), purple vines, or dark-leaved ivies.

Similar varieties: *E. magellanicus* is also startling blue but it is short lived, although it self seeds fairly reliably. Other species (and the closely related *Leymus* or Lyme grass) are often highly invasive.

ERIOPHORUM ANGUSTIFOLIUM
Common cotton grass

Dark, emerald foliage often tinted dark red. Small flower spikes produce striking bunches of cottonlike heads up to 2 inches long. Vigorous spreading habit.
Zones 4–8 H 16 inches x S 24 inches.
Cultivation: Only produces the cotton heads reliably when used in boggy, acidic soils. Sun or light shade.
Uses: Wild pond margins and bog gardens. *Caltha* and *Lysichiton* will add contrast.
Similar varieties: *Eriophorum latifolium* has similar flowers, but clumps more and is tolerant of calcareous soils.

FARGESIA (B)

A genus of tight clump-forming bamboos. The canes are usually strong and upright, with a canopy of of fresh, small leaves, often glaucous.

[1] Hair grass (*Deschampsia cespitosa* 'Goldschleier') is an attractive cultivar for light dry shade and poor soil.

[2] Wavy hair grass (*Deschampsia flexuosa* 'Tatra Gold') really thrives on poor acid soils.

[3] Wild rye (*Elymus magellanicus*) is a lovely blue-gray. Unlike other *Elymus*, it does not run.

[4] Common cotton grass (*Eriophorum angustifolium*) is best grown in a bog garden on acid soils.

[5] *Fargesia nitida* 'Nymphenburg' has a fine arching habit, ideal for a tunnel effect.

[6] Many fescues, like *Festuca glauca,* have fine blue-gray foliage and a tight clump-forming habit.

FARGESIA MURIELIAE
Umbrella bamboo

The most common bamboo in cultivation. Thin green canes turn orange-yellow with age and hold aloft a mass of pale evergreen leaves.
Zones 5–8 H 13 feet x S 3 feet.
Cultivation: Very easy to grow in sites that never dry out.
Uses: As a central specimen, screen or container plant. Underplant with low evergreen cover to keep the canes and wonderful habit in view.
Similar varieties: *Fargesia nitida* is tightly clump forming with dark purple young canes, branching in the second year with many thin, pointed glaucous leaves (H 11 feet x S 3 feet.)

FESTUCA
Fescue

Mostly blue or glaucous green in color and tightly clump forming with threadlike evergreen leaves.

FESTUCA AMETHYSTINA
Sheep's fescue, tufted fescue

Densely tufted blue-green thin foliage. Very noticeable fawn flowers in spring.
Zones 4–9 H 14 inches x S 10 inches.
Cultivation: Full sun in free-draining soil. Annual shearing after flowering keeps them neat. Older plants can be divided at this time if they begin to lose vigor.
Uses: Edging for borders, group planting or containers, with *Bergenia* or neat, low rock garden plants.
Similar varieties: *F. glauca* 'Harz' has turquoise foliage, tinted pink. *F. curvula* subsp. *crassifolia* has a more open habit with steel-blue summer growth.

FESTUCA GLAUCA
Blue fescue, gray fescue

Pale blue, erect threadlike leaves and dense clusters of blue-green flowers turning straw yellow.
Zones 4–9 H 16 inches x S 12 inches.
Cultivation: Full sun in open aspect, not crowded.
Uses: For bed edging with thyme or other creepers. Looks striking when massed in old red clay pots among pebbles and taller container planting.
Similar varieties: There are many to choose from this varied species, and the similar *F. ovina* and *F. valesiaca*. *F. v.* 'Silbersee' is slow-growing, very bright blue, with short, gray-blue flowers with a pincushion habit.
Cultivars: *F. glauca* 'Azurit' has a small habit and brighter blue growth, but is shy to flower.

5

6

GLYCERIA MAXIMA VAR. VARIEGATA

Reed grass, sweet grass

Vertical leaves, heavily striped cream and white with pink flushed new growth. A very colorful and worthwhile plant all year.
Zones 5–10 H 25 feet x S indefinite.
Cultivation: Prefers moist, boggy conditions or even shallow water, but will tolerate good garden soil.
Uses: Deserves to be planted in very large groups by the waterside, but can look equally at home in large containers plunged in shallow ponds.

HAKONECHLOA MACRA

Hakone grass

Slow growing with long, narrow, bamboolike leaves tapering to a point on long stalks arching from tight clumps, maturing to a delicate pink and red in late summer. Deciduous.
Zones 6–9 H 16 inches x S 24 inches.
Cultivation: Requires cool, moist and rich root conditions in well-drained soil, ideally in light shade. Hot sun and dry weather may burn the leaves.
Uses: Very versatile and can be used almost anywhere with a multitude of plants. Good with late-flowering perennials and white variegated hostas.
Cultivars: *H. m.* 'Aureola' has slow-growing clumps of beautiful yellow and pale green striped leaves with an arching habit, red autumn tints, and darker flowers. The leaves of *H. m.* 'Mediovariegata' are finely striped and edged with creamy-white, producing a silvery effect.

HELICTOTRICHON SEMPERVIRENS

Blue oat grass

Arching tufts of narrow, blue-gray leaves. Flowers in midsummer, quickly turning rich golden brown.
Zones 4–9 H 3 feet x S 2 feet.
Cultivation: Well drained, rich soil in sun, not crowded by other plants. Careful pruning of the older evergreen leaves when new shoots appear is preferable to general shearing. Tolerant of very alkaline soils.
Uses: A fine accent plant or for group planting if well spaced. Adds unusual form and color to low perennial borders. Particularly good with silver plants.

x HIBANOBAMUSA TRANQUILLANS (B)

Large veined evergreen leaves taper to a point and can measure 8 inches long by 3 inches wide. The foliage is deep, glossy green, staying fresh through winter.
Zones 6–9 H 8 feet x S 5 feet.
Cultivation: Tolerant of sun or shade in moist but free-draining soil.
Uses: Very useful, suited to screening or specimen use.

HYSTRIX PATULA

Bottlebrush grass

Although normally grown for the spiky, pale silver flowers, which last well into winter, the foliage is also attractive, often tinted red in early spring and again in late summer after flowering.
Zones 4–9 H 36 inches x S 16 inches.
Cultivation: Prefers moist, well-drained soil in part shade. Deciduous.
Uses: Naturalize in woodland gardens or plant around large-leafed bamboos. Looks good with low plants, e.g. cranesbill (*Geranium sanguineum*).

[1] Reed grass (*Glyceria maxima* var. *variegata*) is an aggressive spreader unless it is contained.

[2] Hakone grass (*Hakonechloa macra*) is largely grown for its neat foliage and habit.

[3] Blue oat grass (*Helictotrichon sempervirens*) is a rewarding species that is fairly tolerant of most soils.

[4] Soft rush (*Juncus effusus*) is a common species rarely grown in yards, but with potential for its straight foliage.

IMPERATA CYLINDRICA 'RUBRA'
Japanese blood grass

Thin, almost translucent vertical growth with narrow evergreen leaves. Bright green spring growth gradually changes on the upper half of the foliage to a deep blood red, which stays through winter.
Zones 6–9 H 16 inches x S 8 inches.
CULTIVATION: Difficult to establish unless soil conditions are perfect. Needs rich moisture-retentive soil in full sun where it will spread slowly.
USES: On raised beds, background light will enhance the color and translucence of the foliage. Useful in pots to mix with other plants, especially variegated ones.

INDOCALAMUS LATIFOLIUS (B)

Moderate spreader with thin but strong canes and large evergreen glossy leaves on single branches. Extremely ornamental and tolerant of wind once established.
Zones 6–9 H 6 feet x S 3 feet.
CULTIVATION: Light shade in moist humus-rich soil. Older growth may be thinned to keep open and neat.
USES: Fine when established against a shaded wall or under tall canopied trees. Low planting with *Corydalis* or *Pulmonaria* can protect the soil from drying out around the base. Useful for creating the exotic look.
SIMILAR VARIETIES: *I. solidus* is shorter, but otherwise similar in habit, with tougher and darker green foliage. *I. tessellatus* is very tropical looking with massive plate-sized leaves when mature.

JUNCUS EFFUSUS 'SPIRALIS'
Corkscrew rush

A neat curiosity with semi-evergreen coiled and twisted cylindrical green foliage. The shoots tangle and mat together to form a tight clump.
Zones 5–9 H 16 inches x S 16 inches.
CULTIVATION: Shallow water or boggy soil. Old plants become congested but are easily split and grow freshly.
USES: Unusual in containers with constant moisture.
SIMILAR VARIETIES: The parent species, soft rush (*Juncus effusus*), has straight foliage.

KOELERIA GLAUCA
Blue hairgrass

Similar in habit to blue *Festuca* but stronger in form and leaf size. Porcupine-like tussocks of pale turquoise evergreen leaves arching from a central crown. The pale blue-gray flowers rise just above the foliage in June.
Zones 6–9 H 12 inches x S 8 inches.
CULTIVATION: Tolerant of dry, even chalky, sites in full sun or exposure. Late summer tidying encourages another flush of leaves. Deciduous. Thrives on regular division.
USES: For the dry gravel or rock garden in bold groups. Foreground planting or edging to narrow borders with *Acaena* and dwarf bellflower (*Campanula*).

LAGURUS OVATUS
Hare's tail

A bold and useful annual grass with soft, bushy flower heads of a delicate whitish green.
Zones 6–9 H 24 inches x S 12 inches.
CULTIVATION: Light sandy soil in full sun. Tolerant of drought once established. Seed should be collected and sown *in situ* in succession in spring for summer flowers.
USES: Mostly grown for drying and floral arrangements but useful in a sunny border.

| 1 | 2 |

LUZULA NIVEA
Snowy woodrush

Small growing and highly ornamental evergreen with downy gray-green foliage. Fluffy white flowers in spring toning to a pale rusty cream.
Zones 4–9 H 18 inches x S 12 inches.
Cultivation: Easy and adaptable to a wide range of soils and conditions. Sun or shade and tolerant of some drought when established, even of the difficult root-filled situations around trees. Hard pruning immediately after flowering will provide neat low carpets of the attractive silvery foliage.
Uses: Invaluable as woodland groundcover or as path edging, or as groundcover where almost nothing else will grow. Combine with other low shade-loving plants such as spurge (*Euphorbia amygdaloides* var. 'Robbiae') or varieties of ivy.
Similar varieties: *Luzula lactea* is evergreen, with darker foliage and yellow-white flowers later than *L. nivea*. *L. luzuloides* 'Schneehaschen' is more vigorous, but otherwise identical to *L. nivea*. *L. sylvatica* (greater woodrush) is a versatile evergreen, with dense clumps of long narrow pea-green leaves and plentiful golden-brown spring flowers rising through the foliage (H 20 inches x S 40 inches). *L. s.* 'Aurea' (syn. *L. s.* 'Hohe Tatra') has pale lime-green summer leaves with a burst of deep lemon-yellow winter color. *L. s.* 'Marginata' has finely margined white and green leaves. All are tolerant of dry shade.

MELICA ALTISSIMA 'ATROPURPUREA'
Siberian melic

Soft green arching leaves emerging from neat clumps provide masses of one-sided purple spikes with a papery appearance. Deciduous but reliably perennial.
Zones 4–9 H 36 inches x S 20 inches.
Cultivation: Moist, rich soil in sun or light shade will keep the plant fresh and in flower for ages. Also tolerant of drier conditions.
Uses: Seeds itself, and is pleasing alongside a wide variety of perennials, particularly with cranesbill (*Geranium*), which flower at the same time.

Similar varieties: *M. ciliata* has a profusion of fluffy, off-white panicles that wave in the slightest breeze.

MILIUM EFFUSUM 'AUREUM'
Bowles' golden grass

Golden-yellow papery leaves hold their color even in shade. The nodding flowers are delicate pale yellow,
Zones 6–9 H 12 inches x S 24 inches.
Cultivation: Tolerant of sun or shade, including dry shade, though more moisture is required when it is grown in hot areas. Slowly spreading clumps will usually multiply by self seeding.
Uses: Brightens up the dullest corner with an almost fluorescent display. Works particularly well with spring perennials, especially those with blue or purple flowers, which form a pleasing contrast with the yellow foliage.

MISCANTHUS
Silver grass, eulalia

Tall graceful foliage and late summer flowers/seed heads that last well into winter, providing almost nine months of striking elegance. There are a great many *Miscanthus sinensis* cultivars, many of which appear similar as young plants but mature with subtle differences of flower color, foliage, shape, and habit.

MISCANTHUS SACCHARIFLORUS
Silver banner grass, Amur silver grass

Tall, with a subtropical appearance. Broad clumps of bamboolike canes with long arching strappy leaves. Small silver flowers.
Zones 4–9 H 8 feet x S 5 feet.
Cultivation: Moist fertile soil in full sun. Withstands strong winds even in the winter, when it pales to a rich straw color.
Uses: Screening or for specimen use as a feature plant or as background planting.
Similar varieties: *M. floridulus* (giant miscanthus) is slightly more reliably evergreen than *M. sacchariflorus*, and it has better display of late coloring.

[1] Snowy woodrush (*Luzula nivea*) has fine white hairs on the outermost edges of its leaves which contribute to its interesting and unusual appearance.

[2] Bowles' golden grass (*Milium effusum* 'Aureum') is a useful plant to have in the garden because it is one of the minority of grasses whose best season of interest is the early summer.

[3] *Miscanthus sinensis* 'Malepartus' is one of the best of the larger silver grass (*Miscanthus*) cultivars.

[4] *Miscanthus sinensis* 'Morning Light' has cream edging to its leaves.

[5] The delicate flowers of *Miscanthus x oligonensis* 'Zwergelefant' emerge on the stem.

MISCANTHUS SINENSIS
Japanese silver grass, eulalia

Tightly clumping with rigid, vertical stems of long narrow foliage. Autumn tints vary from pale, creamy-yellow through a pink tan to deep orange or purple-red. Flowers on some cultivars appear as early as July and a few as late as October. The whole plant dies back by midwinter. Zones 4–9 H 3–6 feet x S 3–4 feet.

Cultivation: Best in rich, moisture-retentive soil but also adaptable to drier sites, which give a shorter habit and earlier flowering. Full sun provides the best flower and foliage color. Narrow-leaved forms seem more drought tolerant. Hard prune in spring before new growth and occasionally divide (every 5–7 years).

Uses: As screens, focal points, or in mixed border planting or as pond edging (but not in boggy soil). Look best accompanied by large late-flowering perennials.

Cultivars: *M. s.* 'Kleine Fontäne' has bold red arching flowers from mid to late summer, turning silver and fluffy, with strong vertical stems and red-tinted foliage late in season (H 6 feet x S 20 inches). *M. s.* 'China' is similar to *M. s.* 'Kleine Fontäne but slightly later. *M. s.* 'Flamingo' has deep claret-red plumes turning pink to silver, narrow foliage and an arching habit. It is strong in the winter (H 5 feet). *M. s.* 'Gracillimus' (maiden grass) is an old cultivar with very fine-textured foliage with silvery midribs. It is pale in growth until autumn, when it will color with yellow and orange. The flowers are very late and only appear after a long growing season (H 4 feet). *M. s.* 'Nippon' is similar to *M. s.* 'Gracillimus' with copper autumn tones. *M. s.* 'Graziella' has large silver-white plumes that grow high above the foliage, with good autumn color in colder climates. It is quick to establish (H 4 feet). *M. s.* 'Undine' is similar and also easy to establish, but bigger at 6 feet. *M. s.* 'Kaskade' has deep purple-brown flowers turning silvery-pink with orange-red autumn leaf color and an elegant arching habit (H 6 feet). *M. s.* 'Kleine Silberspinne' is a compact plant with very narrow silvery leaves. Dark red-brown plumes like narrow fans appear in midsummer. It is very sturdy and lasts longer than most grasses through winter. Ideal for the smaller garden (H 3 feet). *M. s.* 'Silberspinne' is only a little taller than *M. s.* 'Kleine Silberspinne' with a slightly more open habit. *M. s.* 'Malepartus' has large deep red flowers that are slightly arching, strong stems, and broad leaves that also turn a rich purple-red (H 8 feet). *M. s.* 'Rotsilber' is similar. *M. s.* 'Morning Light' is slender and similar to *M. s.* 'Gracillimus.' It is shy to flower, with pale leaves delicately edged white for a shimmering effect (H 3 feet). *M. s.* 'Silberfeder' is very tall with arching pink plumes, quickly turning to silver. The leaves have a broad silvery-white central vein and good yellow autumn color. It has a strong habit, lasting well into the spring (H 10 feet or more). *M. s.* 'Strictus' (porcupine grass) has a tight vertical habit with green leaves horizontally banded yellow. It flowers late, but the foliage is tough and almost semi-evergreen, giving late color (H 6 feet). *M. s.* 'Variegatus'has broad creamy-white edged leaves and shy red flowers (H 6 feet). *M. s.* 'Zebrinus' (zebra grass) is identical in coloring but more open in habit, with foliage arching at the top (H 10 feet). *M. s.* 'Pünktchen' is a new dwarfer form with the same yellow banding and more reliable, deep red flowers (H 4 feet). *M. s.* 'Yakushima Dwarf' is a tiny form, as the name suggests, growing to only 1 foot with sporadic dark brown flowers above the foliage. It forms curious hummocks, with leaves curling at the tips.

Similar varieties: *M. x oligonensis* 'Zwergelefant' has wide leaves and grows to 6 feet. The name means dwarf elephant, referring to the shape of the young flowers.

MOLINIA CAERULEA

Moor grass

Tufts of narrow foliage and flowers form on straight
stems from late summer on. All forms are
deciduous, and sub-species *arundinacea* types lose their
seed heads in early winter.
Zones 4–9 H 3 feet x S 2 feet.

CULTIVATION: Moist fertile soil in full sun provides the best
results. Very alkaline soils are not tolerated. Clear messy
growth in early spring, and divide when clumps start to
grow thinly from the crown.

USES: For single specimen accent planting or grouping
with other grasses and perennials. The usual bright
yellow or orange autumn coloring associates well with
other autumn foliage.

CULTIVARS: *M. c.* subsp. *arundinacea* 'Bergfreund' has
green strappy leaves, almost vertical in habit, with tall
dark purple flowers from late July. Autumn color is pale
yellow (H 4 feet x S 2 feet). *M. c.* subsp. *arundinacea*
'Transparent' has strong foliage, and delicate airy flowers
on large panicles do give an almost transparent effect,
earning it its name. Late summer and autumn foliage is
bright yellow, with the stems standing into winter
(H 6 feet x S 2 feet). *M. c.* subsp. *caerulea* 'Windspiel' has
large panicles of golden flowers held high above glossy
green, straplike leaves. This is another good autumn gold
(H 6 feet x S 20 inches). *M. c.* subsp. *caerulea* 'Edith
Dudszus' is a shorter form with neat, spiky leaves and
dark red stems producing late matt-purple flowers. It has
pale creamy-yellow color in autumn (H 3 feet x S 1 foot).
M. c. subsp. *caerulea* 'Heidebraut' has broad fountains of
narrow foliage and short dark stems of pale flowers. A
deep orange-yellow, it turns dark russet-red with the first
frosts (H 40 inches x S 16 inches). *M. c.* subsp. *caerulea*
'Moorhexe' is much smaller in stature, forming a tight
clump of gray-green pointed leaves. Dark flowers are
often produced as early as July (H 30 inches x S 12
inches). *M. c.* subsp. *caerulea* 'Variegata' is a variegated
form with cream and green striped leaves. Very dark
flowers appear in July on thin, shiny yellow stems.

MUHLENBERGIA JAPONICA 'CREAM DELIGHT'

Muhly grass

Colorful mat of creamy foliage. Thin white striped leaves
and stems turn buff-yellow in autumn just after the pale
silvery flowers appear.
Zones 5–9 H 8 inches x S 36 inches.

CULTIVATION: Winter clearing and hard spring pruning with
division every few years are necessary to keep the plant
young and fresh. Sun and average soil.

USES: As a mat around taller grasses such asmoor grass
(*Molinia*) or silver grass (*Miscanthus*) or in rock gardens.

PANICUM CLANDESTINUM

Deer tongue grass

Distinct bright, almost bamboolike foliage. Soft green
leaves, broad and tapering sharply to a point from the
reddish stems. Fine russet autumn hues. Compact.
Zones 4–9 H 36 inches x S 30 inches.

CULTIVATION: Moist but well-drained soil in light shade.

USES: As woodland path edging or groundcover by pools.

PANICUM VIRGATUM

Switch grass

Clump-forming and upright with long narrow foliage and
sprays of tiny silvery or red flowers in loose panicles in
late summer.
Zones 5–9 H 36–40 inches x S 16–20 inches.

CULTIVATION: Moisture-retentive soil in sun is ideal, but all
adapt to poorer conditions once established. Good in
wind and salt spray.

USES: Adds character and height to perennial borders
and mixes well with other late-season grasses such as
feather grass (*Stipa*) or silver grass (*Miscanthus*), and late
perennials like stonecrop (*Sedum spectabile*).

CULTIVARS: *Panicum virgatum* 'Hänse Herms' has deep
purple flowers and plum-tinted foliage (H 36 inches x
S 16 inches). *P. v.* 'Rehbraun' is similar but not so good.

P. v. 'Heavy Metal' has metallic-blue vertical leaves that turn pale yellow in autumn and late yellow-brown flower panicles (H 4 feet). *P. v.* 'Rotstrahlbusch' has deep claret-red autumn color and dark greeny-red flowers (H 36 inches x S 18 inches).

PENNISETUM ALOPECUROIDES
Fountain grass

Showy deciduous grass forming tight clumps of stiff pointed foliage with many large, pink tinted, foxtaillike fluffy flowers/seed heads from midsummer. Autumn foliage tints vary from rich creamy-yellow to pink, depending on the site.
Zones 6–9 H 40 inches x S 24 inches.
CULTIVATION: Tolerant of most soils, but dislikes winter wet. Full sun or very light shade.

USES: Good autumn and winter effect, the late colors going well with the autumn yellow and red of many shrubs, as well as the yellow and purple-violet of late perennials. Can be used as a focal point or in groups.
CULTIVARS: *P. a.* 'Hameln' is dwarf with short rigid flowers (H 20 inches x S 12 inches). *P. a.* 'Woodside' is compact with a reliable display of flowers. It tolerates more shade than other cultivars (H 30 inches x S 15 inches.)
SIMILAR VARIETIES: *P. orientale* (Oriental fountain grass) has pink flowers fading to white (H 24 inches x S 20 inches).

PHALARIS ARUNDINACEA
Ribbon grass, gardeners' garters

Tall grass with bamboolike foliage, bright glaucous green, paling to warm buff-yellow. Flowers are tall, pink-tinged panicles. Strongly spreading.
Zones 4–9 H 40 inches x S 30 inches.
CULTIVATION: Very moist soils in sun or light shade; wind and exposure tolerant.
USES: A bold pond-edger tolerating boglike conditions. Suitable only for a larger yard unless thinned. Use with giant rhubarb (*Gunnera*) for a controlled wilderness.
SIMILAR VARIETIES: *Phalaris arundinacea* var. *picta* 'Feesey' is the form most suitable for a smal yard, with an almost completely pure white effect.

[1] Moor grass (*Molinia caerulea*) is inconspicuous until it flowers.

[2] Switch grass (*Panicum virgatum* 'Rehbraun') has reddish autumn tints.

[3] Fountain grass (*Pennisetum alopecuroides*) is good for late summer and autumn interest.

[4] Oriental fountain grass (*Pennisetum orientale*) has violet flowers that turn to straw-colored seed heads.

[5] Ribbon grass (*Phalaris arundinacea* var. *picta* 'Feesey') is a moisture-lover with exotic variegation.

PHYLLOSTACHYS (B)

An important and remarkable genus with tall, often thick canes with a groove (sulcus) on the sides. New canes (culms) develop taller and thicker as the rhizome below ground matures. The genus has many species, variable in habit. Many are invasive in their natural habitat but are more refined and compact in cooler temperate regions.

PHYLLOSTACHYS AUREA
Fishpole bamboo

Tall, almost vertical habit, foliage held high. Compact unless conditions are wet and warm.
Zones 5–9 H 16 feet x S 5 feet.
Cultivation: Rich, moist soil. Occasional thinning of older canes will result in stronger growth.
Uses: Ideal specimen plant or screen. Suitable for small yards due to its vertical habit.

PHYLLOSTACHYS AUREOSULCATA
Crookstem bamboo, yellow grove bamboo

One of the hardiest. Vertical green canes with yellow grooves, often zigzagging at the base. Small deep green foliage is always fresh and glossy. A moderate runner.
Zones 5–9 H 16 feet x S 8 feet.
Cultivation: Occasional thinning necessary to encourage the display of young canes.
Uses: As edge or pathside planting to appreciate the unusual canes, while large mature groves can be thinned and paths cut through. Heavy stepping stones will divert young shoots and create a narrow bamboo tunnel. The canes of this species and the its golden cultivars should not be congested with intrusive planting or the vertical effect will be lost. Also makes a useful screen or hedge, even in exposed locations.

PHYLLOSTACHYS BAMBUSOIDES
Japanese timber bamboo

Large clumps of very strong, thick canes of rich shiny green. Medium-sized pointed leaves. Tight in habit until mature, when it forms an open grove of vertical canes.
Zones 5–9 H 20 feet x S 8 feet.
Cultivation: Very adaptable in acid or alkaline moisture-retentive soils. In cool regions it is best planted in sun to ripen the late emerging shoots. Older growth needs removing regularly. Can be rampant in warmer areas.
Uses: Ideal for screening or for a bold central feature. In larger yards where large island beds can be created, Mahonia is a good associate.

PHYLLOSTACHYS BISSETII

Rich, glossy, purple-green new canes and dense foliage quickly form a colony. Older canes turn pale green.
Zones 5–10 H 18 feet x S 10 feet.
Cultivation: Any moisture-retentive soil.
Uses: As a screen, hedge, or background. Will bind loose banks or the edges of natural ponds. Strands of willows, e.g. *Salix alba* 'Britzensis,' with bright winter stems, positively shine against the dark foliage.

PHYLLOSTACHYS NIGRA
Black bamboo

Shiny, jet-black canes form tall neat clumps. The small pale green leaves enhance the dark canes. Young canes, green at first, turn black over a two- or three-year period.
Zones 5–9 H 16 feet x S 5 feet.

Cultivation: Full sun ripens the canes best, but the plants must not dry out. Pruning the lower branches will also allow light to the canes and help ripening.
Uses: Very adaptable and can be used successfully in large containers on patios or balconies. The ideal focus for a small or medium yard, the unusual black coloring is ideally suited for association with many climbers, such as the ivy *Hedera helix* 'Silver King.' Blue or white flowering perennialsor shrubs or dark purple foliage are also very effective companions.

PHYLLOSTACHYS VIVAX
Smooth-sheathed bamboo

Tall and stately with very thick pale green canes and a high foliage canopy. Very compact at the base and broad at the top, forming a distinct V-shape. The new culms are striking with dark spotting.
Zones 5–9 H 20 feet x S 5 feet.
Cultivation: Young plants will take time to mature and few canes are initially produced until the underground rhizomes develop, the initial growth being thin and juvenile (this is common in many of the taller cane bamboos). Large canes will eventually, after a number of years, supersede the early growth, which can then be removed. Sun or shade.
Uses: A stunning specimen or feature plant. Excellent weeping over water. Looks good with large-leaved companion plants.

PLEIOBLASTUS (B)

A genus of huge variety ranging from the smallest bamboo (*Pleioblastus pygmaeus*) to taller leafy plants. The shorter species are valuable for garden use because of the many colorful foliage forms, being variegated with yellow, cream, or white. Although evergreen and woody, many *Pleioblastus*, especially the shorter species and cultivars, generate a mass of annual growth from the base and are often seen at their best when all previous years' material has been removed. Most are spreading in habit, but with occasional control the rhizomes can be checked to form a compact feature. Nearly all the species can be used successfully for container gardening.

PLEIOBLASTUS AURICOMUS
(SYN. *P. VIRIDISTRIATUS*)

Widely cultivated yellow-foliage bamboo. The long pointed pale golden leaves are variably striped green. Moderate in habit, running only when mature and moist.
Zones 5–9 H 5 feet x S 5 feet.
Cultivation: Once established, late winter shearing or thinning will allow new colorful growth to dominate. Full sun always provides the best color.
Uses: Underplanting for taller bamboos and shrubs. Pond side or path edging where the color is best appreciated. Very versatile; not out of place in a perennial border.
Similar varieties: *P. variegatus* is a high-quality variegated plant (H 40 inches); best if cut hard back occasionally.

PLEIOBLASTUS PYGMAEUS

Tiny pointed blue-green leaves on short wiry canes. Some of the leaves bleach completely in winter.
Zones 5–9 H 12 inches x S 40 inches or more.
Cultivation: Thrives on hard pruning in late winter, or the habit is loose and leggy. Vigorous and invasive, but easily controlled with a sharp spade. Sun or half shade.
Uses: Path and drive edging or very low underplanting for taller plants. Very resilient in containers.

[1] The dark foliage of crookstem bamboo (*Phyllostachys aureosulcata* 'Aureocaulis') contrasts with the stems.

[2] *Phyllostachys bambusoides* 'Castillonis' is a yellow-stemmed cultivar of Japanese timber bamboo.

[3] *Phyllostachys bissetii* is vigorous once established, making it suitable for screening.

[4] *Phyllostachys nigra* is the extremely popular black-stemmed bamboo.

[5] Smooth-sheathed bamboo (*Phyllostachys vivax*) is a neat, upright grower with a mass of thick canes.

[6] *Pleioblastus variegatus* has a reputation as one of the finest of variegated plants.

PLEIOBLASTUS SHIBUYANUS 'TSUBOI'

One of the brightest of bamboos, with pale green canes holding bunches of cream-suffused leaves. Extremely vigorous, forming a dense grove.
Zones 5–9 H 6 feet x S 6 feet.
Cultivation: Tolerant of a wide range of soils in sun or part shade.
Uses: The overall cream coloration of the foliage is best appreciated if the plant is not crowded out, so shorter foreground planting should be kept simple—try elephant's ears (*Bergenia*) or short woody plants. Good when confined in pots.

POA CHAIXII
Broad-leaved meadow grass

Broad and flat dark green evergreen leaves make a neat clump that is topped by a froth of greenish flowers during May and June.
Zones 5–10 H 30 inches x S 16 inches.
Cultivation: Ideal for shade and even tolerant of dry soils once established. Shear hard after flowering to generate fresh foliage.
Uses: Very airy and light in flower, so it is ideal for wood-land path edging or for surrounding hard landscape materials to soften the edges. Its tolerance of shade makes it an indispensable garden plant.

[1] Arrow bamboo (*Pseudosasa japonica* 'Tsutsumiana') has highly ornamental swollen canes.

[2] The bamboo *Sasa kurilensis* 'Shimofuri' has distinctive banded leaves.

[3] *Sasa kurilensis*, from the Kurile Islands between Japan and Siberia, is possibly the hardiest of all bamboos.

[4] *Sasa palmata* f. *nebulosa* is one of the more invasive bamboos, but highly effective given space, or a container to limit its growth.

[5] The bamboo *Sasaella masamuneana* f. *albostriata* has attractive variegated foliage and prefers sun.

PSEUDOSASA JAPONICA (B)

Arrow bamboo

Robust and vigorous species with many dark olive-green canes. Leaves are thick large and glossy green, paler beneath. Forms a bold, vertical plant very quickly. Zones 6–9 H 16 feet x S 6 feet.

CULTIVATION: Tolerant of exposure and useful in sun or shade, this is one of the easiest bamboos to establish. Runners are easy to control.

USES: Often used as a windbreak or screen where its performance is exemplary. Also good as a lone specimen or part of a bold planting with lower plants.

SACCHARUM RAVENNAE

Sugar grass, plum grass

Dense tussocks of hairy gray-green strappy foliage support large silver-purple feathery plumes on tall stems. Deciduous in cold climates.
Zones 7–10 H 6 feet (in flower) x S 30 inches.

CULTIVATION: Full sun and warmth will produce a mass of flowers. Good winter drainage is essential if the plant is to survive. Clear up annually in spring.

USES: A tall plant for the dry gravel garden and also useful in coastal regions. Can be used in a California-style garden combined with gray-leaved plants or yucca.

SASA (B)

A distinct genus usually having large broad leaves on thin but strong canes, which often bleach on the edges during the winter, creating a variegated effect from afar. Many species provide an instant subtropical appearance.

Most are quickly spreading, but can be successfully confined in pots making them suitable for any yard.

SASA KURILENSIS

Incredibly tough species. Tall sheathed canes eventually turning pale yellow-green hold thick leathery leaves, which bleach slightly in winter. It provides a feeling of formality and is evenly vigorous, forming a stately grove. Zones 5–9 H 6 feet x S 5 feet.

CULTIVATION: Sun or shade and very tolerant of exposure.

USES: The modern approach is shady woodland edge with dogwoods. Waterside planting with ferns and mossy rocks creates a prehistoric touch.

CULTIVARS: *S. k.* 'Shimofuri' has white longitudinal bands running along the broad leaves.

SASA PALMATA F. NEBULOSA

Beautiful form with fresh green canes forming dark brown blotches in sun with age. The leaves are the most exotically large of the *Sasa* species. Very vigorous, but controllable with regular digging or the use of a barrier. Zones 5–9 H 8 feet x S 8 feet.

CULTIVATION: Regular thinning of old canes and foliage as new growth starts in spring will keep the plant fresh and glossy. Tolerant of most soils in sun or shade.

USES: For broad screening, pondside, or even in large containers. Wall planting is very effective when the leaves are striking against red brick or whitewash.

SASAELLA MASAMUNEANA (B)

Similar to *Sasa* with broad, but smaller leaves. This species has a profusion of dark purple-black thin canes with a fresh display of foliage well above the ground. Spreads evenly.
Zones 6–9 H 6 feet x S 5 feet.

CULTIVATION: Can be sheared annually; otherwise, thin old material as new spring growth starts. Tolerant of acid or alkaline soils; coloring better in full sun.

USES: Groundcover or underplanting for taller bamboos. The dark canes associate well with variegated grasses such as ribbon grass (*Phalaris*) or velvet grass (*Holcus*).

CULTIVARS: *Sasaella masamuneana* f. *albostriata* has good variegated foliage and a moderately spreading habit.

SCHIZACHYRIUM SCOPARIUM

(SYN: ANDROPOGON SCOPARIUS)

Little bluestem

An attractive prairie grass, invaluable for later season effect. Clumps of pale green, slightly hairy foliage, change to plum-purple as the flowers develop in late summer. Flowers are pale blue-green, rapidly turning to seed as the whole plant ages to bronze and orange with the first frosts.

Zones 3–10 H 3 feet x S 2 feet.

Cultivation: Versatile, thriving on a wide range of soil types and easy to establish in full sun, where it will tolerate drought. Will seed to form broad colonies, but easily controlled by hoeing. Shear annually in spring.

Uses: Massed planting or groundcover with spot planting of silver grass (*Miscanthus*) or other tall grasses. The autumn color is especially good with the foliage of many of the new alum root (*Heuchera*) hybrids.

SCHOENOPLECTUS LACUSTRIS SUBSP. TABERNAEMONTANI 'ZEBRINUS'

Zebra rush

Cylindrical rushlike leaves horizontally banded with creamy yellow, which is often more dominant than the green. The flowers are insignificant.

Zones 7–9 H 3 feet x S 3 feet.

Cultivation: Needs rich, moist, or wet soil in sun and tolerates marginal conditions up to 6 inches deep. Needs tidying during late winter.

Uses: Poolside, for reflection in the water. Superb with night lighting since the transparency of the cream banding highlights the variegated effect.

[1] A sheath of the common, but magnificent, *Semiarundinaria fastuosa*.

[2] Prairie cord grass (*Spartina pectinata* 'Aureomarginata') is good for large expanses of waterside, where it can safely spread.

SEMIARUNDINARIA FASTUOSA (B)

A moderate runner with very stiff, erect canes, usually extremely thick, turning rich plum-purple with silver streaks as they age. The sparse fresh green foliage is dense, held on short branches close to the canes. Distinct in habit, and very hardy.

Zones 5–10 H 20 feet x S 8 feet.

Cultivation: Easy to establish in soils rich in organic matter. Sun or light shade. Prune old, thin shoots as the thicker ones emerge. Tolerant of wind, exposure, and coastal conditions.

Uses: Can easily be maintained as a tight clump in a smaller yard as a solitary specimen, not congested and cluttered by other plants.

SESLERIA CAERULEA

Blue moor grass

A short evergreen grass with unusual two-tone coloring. The short rigid upright leaves are dark and shiny on the upper surface and pale matt-blue on the lower side. The early spring flowers emerge purple and gradually turn straw yellow.

Zones 5–9 H 1 foot x S 1 foot.

Cultivation: Very tolerant of alkaline soils as long as there is good moisture with some drainage. Best in full sun or very light shade.

Uses: For low massed cover or edging. Excellent with dwarf spring bulbs or mixing with other short grasses.

SHIBATAEA KUMASASA (B)

Short habit with thin zigzag canes. Very distinct, with dense mounded evergreen foliage bleaching at the tips during winter. Leaves are small and diamond shaped.

Zones 5–9 H 4 feet x S 20 inches.

Cultivation: Prefers cool, moist soil and is usually clump forming unless conditions are very warm and wet.

Uses: A fine container plant if never allowed to dry out. Feature planting by paths or patio.

Similar varieties: *S. lancifolia* has larger leaves on more noticeable stalks, and is flatter in habit.

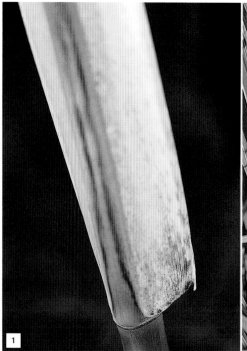

SORGHASTRUM AVENACEUM

(SYN: CHRYSOPOGON AVENACEUS)

Indian grass, gold beard grass

Upright and clump-forming prairie grass with pale blue-green arching leaves. The late summer flowers are majestic shiny bronze and dry well. The foliage turns through rich purple to orange-yellow in autumn.
Zones 4–9 H 40 inches x S 16 inches.
CULTIVATION: Drought resistant and hardy, growing taller with more moisture. Adaptable to varying soil types and strong in the wind. Shear annually in late winter.
USES: Natural planting with other prairie grasses or as a feature in the grass or perennial border.

SPARTINA PECTINATA 'AUREOMARGINATA'

Prairie cord grass

Long, arching pale green leaves have golden margins. The flowers are not ornamental; remove them to add vigor to the leaves, which turns vivid amber in autumn and winter.
Zones 4–9 H 6 feet x 3 feet.
CULTIVATION: Good, moist fertile soil is essential or growth will be weak. Full sun sharpens the contrast of gold leaf margins against the green. Prune hard in early spring. Tolerant of exposure. Can be invasive.
USES: Waterside planting with giant rhubarb (*Gunnera*), where it will form thickets of foliage, good for wildlife.

SPODIOPOGON SIBIRICUS

Frost grass

Highly ornamental. Stout stems have angled foliage drooping to the horizontal, giving a bamboolike effect, Broad panicles of green-purple flowers appear in late summer. Deep rusty-brown autumn coloring.
Zones 5–9 H 40 inches x S 16 inches.
CULTIVATION: Moist fertile soil, sun or light shade. Tolerant of exposure. Divide every 3 or 4 years, prune in spring.

USES: Especially useful in the late-flowering perennial bed as a foreground accent or in the winter garden outlined in frost. The architectural qualities of this grass should be highlighted and additional planting kept simple, e.g. low-growing perennials like elephant's ears (*Bergenia*).

SPOROBOLUS HETEROLEPSIS

Prairie dropseed

Arching clumps of long, hairlike, deep green leaves. Long, scented, delicate flower panicles quickly turn to seed and droop over the foliage with their weight. Turns delicious pale chocolate-brown in autumn and winter.
Zones 3–9 H 24 inches x S 30 inches.
CULTIVATION: Established plants are very tolerant of dry arid conditions and hot sun where they show their true color. Annual thinning and regular division will enhance the weeping appearance.
USES: Raised beds or dry gravel gardens, or alongside paths to benefit from the scent of the flowers.

STIPA

Feather grass, needle grass

Large group of highly ornamental grasses, most tolerant of hot, dry climates. Grown for the flower/seed heads.

STIPA ARUNDINACEA

Feather grass

Elegant, long and narrow leaves on fountainlike clumps. Foliage turns rusty orange as summer closes. Feathery flowers arch almost to the ground and fall quickly, leaving a fiery display of foliage until the next spring.
Zones 6–10 H 20 inches x S 24 inches.
CULTIVATION: Tolerant of drought and bright sun once established. Can self-seed.
USES: Highly effective in winter combined with golden variegated evergreens or adding color to containers, borders or gravel gardens.

[3] Frost grass (*Spodiopogon sibiricus*) has almost bamboolike foliage.

[4] Prairie dropseed (*Sporobolus heterolepis*) is an important element of old prairies; this species is reasonably drought tolerant.

STIPA BARBATA

Exceptional species with immensely long featherlike seed heads that move in the slightest breeze.
Zones 6–10 H 30 inches x S 24 inches.
CULTIVATION: Best in dry, sunny situations. Dislikes close competition from other plants.
USES: With other dry-land, rock-garden and subtropical plants, preferably lower growing.
SIMILAR VARIETIES: *S. pennata* is similar in habit, but has shorter seed head feathers.

STIPA GIGANTEA

Giant feather grass

Worthy of a place in any yard. The long arching gray-green foliage supports the tall flower spikes, which are produced late in the spring and last virtually all year.
Zones 6–10 H 8 feet x S 3 feet.
CULTIVATION: Restrict clearing to the autumn or winter as spring pruning will often damage the newly emerging flowers. Split large clumps every 4 or 5 years or flower production will be dramatically reduced; this is best done immediately after flowering. Discard the old central crowns and replant the fresh juvenile growth from the edges of the clump. Tolerant of most types of soils and conditions including dry and exposed sites, as well as coastal gardens.
USES: Too good to hide, so give a prominent placing. The foliage is low and dour, so brighten it up with low plant-ing of more colorful, lower-growing plants.

STIPA TENUIFOLIA

Often called *S. tenuissima*. Tufts of very fine leaves and silky seed heads forming cloudlike clumps.
Zones 6–10 H 16 inches x S 8 inches.
CULTIVATION: Any reasonable soil in sun. Short lived but self seeds well.
USES: Lovely *en masse*, and also when mixed with low-growing dry habitat plants. Effective when self-sown in the cracks of paving.
SIMILAR VARIETIES: *S. capillata* is taller (36 inches) and coarser but its flower/seed heads are similarly hazy.

UNCINIA RUBRA

Hook sedge

Deep red-brown glossy leaves arch from tight clumps.
In harsh winters, the leaves may tint with orange and
copper-red. Flower/seed heads are insignificant.
Zones 7–9 H 18 inches x S 20 inches.
Cultivation: Full sun for best color in moisture-retentive
soils. Regular division or thinning of old growth is vital.
Uses: An invaluable color in the yard. Very attractive with
the winter color of greather woodrush (*Luzula sylvatica*
'Aurea') and the sedge *Carex* 'Frosted Curls.'
Similar varieties: *Uncinia clavata* has evergreen tufts of
gray-green foliage. Large club-shaped yellowish flowers
turn dark brown on protruding spikes.

YUSHANIA ANCEPS (B)

A bold, running species forming large colonies of upright
fresh green canes with billowing plumes of foliage
arching down and out from the tall canes. Its beauty is
best appreciated from a distance.
Zones 6–9 H 20 feet x S 8 feet.
Cultivation: Thin old canes to produce colonies of
well-spaced younger and fresher growth. Sun or shade
in almost any reasonable soil.
Uses: Large landscape feature or distinct planting in
parkland. Makes excellent hedging, screening, and
arching walkways.
Similar varieties: *Y. maling* forms a grove in time, with
tall (up to 12 feet) arching stems.

[1] *Stipa barbata* is only really successful on
dry soils with little competition.

[2] Giant feather grass (*Stipa gigantea*) is
one of the most rewarding grasses and it
establishes itself very quickly.

[3] *Stipa tenuifolia* has very fine flower and seed
heads that result in a cloud effect.

[4] Hook sedge (*Uncinia rubra*) has rich mahogany
brown leaves, but needs moist soil to thrive.

[5] Masses of small leaves are a distinctive
feature of *Yushania anceps*.

PRACTICALITIES

ONE OF THE WONDERFUL THINGS ABOUT
GRASSES IS HOW TOLERANT THEY ARE OF
A WIDE RANGE OF CONDITIONS. ANOTHER
IS HOW LITTLE MAINTENANCE THEY NEED.
NEVERTHELESS, THIS DOES NOT MEAN
THAT THEY WILL GROW EVERYWHERE AND
ANYWHERE, OR THAT THEY ARE TOTALLY
PROBLEM-FREE. AS WITH ANY PLANTS, A
BASIC UNDERSTANDING OF THEIR NEEDS
AND HABITS WILL GO A LONG WAY.

OPPOSITE: Bamboos are definitely more fussy about
growing conditions than most grasses; shelter
from wind and a well-drained but moist soil are
essential. This is *Phyllostachys violescens*.

ABOVE: Sedges, on the other hand, are very
tolerant, especially of poor soils. This is
Carex sylvatica, which does well in light shade.

SELECTING GRASSES FOR YOUR GARDEN

How suitable a plant is for its chosen site really ought to be the most important criterion when selecting a plant. However striking it ought to look, if conditions are not right, then it will not thrive, and there is no point in planting it. Fortunately, grasses are remarkably tolerant plants, with lack of light being the chief limiting factor on what varieties can be grown. Any site that receives sun for less than three-quarters of the day during the growing season is likely to be less than satisfactory for the majority of grasses; for more shady situations, those varieties that have adapted themselves for life in reduced light will need to be chosen.

Although certain species of grasses tend naturally to be found on alkaline soils, and others on acid soils, this does not mean that they have to be rigidly grown only on these soils, as is the case with many other cultivated plants. Grasses are tolerant of a wide variety of soil types, with only the extremes likely to cause problems. Most true grasses grow much better on fertile soils, but they flourish well enough on poorer ones, with the tough and tolerant sedges tending to be particularly successful on the latter. Consequently, there is no need for the extensive "soil improvement" that traditional gardening has advocated (see Planting page 105).

After light levels, soil moisture is probably the most important factor to consider. Soil that regularly suffers waterlogging should be planted only with those species known to be able to survive this condition—others may rot. The same is true of soils that are regularly affected by drought; although most grasses will survive, only those from dry environments will really flourish.

Other practical considerations to bear in mind when choosing grasses are mainly to do with habit of growth. As with all plants, you should take care to ascertain the eventual height of a plant before buying. Plants in pots in garden centers are never the height they are capable of growing in the open ground. Spread is often of greater importance than height; this should be checked, too. It is worth bearing in mind that some of the more popular grasses have a great many varieties in various sizes. If you adore Japanese silver grass (*Miscanthus sinensis*), for example, but feel that at six feet the popular variety 'Silberfeder' is too large, there are many other cultivars that are smaller, although they might take a little more searching out from specialized nurseries.

GRASSES THAT RUN

Since some grasses are notorious for spreading rapidly, those particular species should be planted only where there is space for them. (These are indicated in the Directory, *see pages* 78–99.) Most spreaders are also moisture-lovers, and in any case species that run are more likely to spread on wetter soils. An exception are species of wild rye (*Elymus*), which are dry-habitat plants; they tend to run more in light, sandy soils.

Certain bamboos are notorious runners, as can sometimes be appreciated in old yards where they are seen to cover large areas of ground. Most bamboos run more rapidly in warm and moist climates; species that are quite restrained and slow growing in cooler areas grow at a tremendous pace. In regions that experience very warm and humid summers, great care should be taken in selecting species and in preventing their running into neighboring property. In such climates it may be best to stick to those that are known never to run, such as *Chusquea culeou*.

Rampant species of grasses, including bamboos, may be restrained by container culture, or by growing in bottomless containers, so that the plant's main roots are in contact with the soil below them, but their horizontal running roots are prevented from spreading. Any solid barrier running down to a foot or so below the soil surface is adequate to prevent running.

CHOOSING AND SITING BAMBOOS

Bamboos have rather more specialized requirements than other grasses, needing an environment that is sheltered from strong and cold winds, both of which cause leaf scorch, and which offers consistently moist growing conditions, on soil that is well drained but never dries out. Positions in full sun should be avoided, especially outside cool temperate regions. Light woodland, slopes with water seepage, and valley bottoms are the situations where the best bamboos are seen.

Bamboos do vary in hardiness, but any responsible nursery will advise on species that may not be hardy. It would be advisable to ask about unfamiliar species, especially if you live in an particularly cold area.

RIGHT: Fountain grass (*Pennisetum alopecuroides*) is one of a genus that is very beautiful, but intolerant of winter damp, tending to grow more easily in midcontinental climates, with hot summers and cold, dry winters, than in capricious maritime climates.

BELOW: Low-growing varieties of bamboo can be clipped and maintained as edging plants, as traditionally done in Japan.

BUYING PLANTS

Most garden centers now sell a basic selection of grasses, with specialized perennial and grass nurseries selling a great many more. Bamboo, being rather a cult plant, tends to be found in greatest variety in nurseries that sell nothing but bamboo.

Specialized nurseries may be located through plant directories, through advertisements in more serious gardening magazines, and increasingly commonly, on the Internet. Many, indeed most, provide a mail-order service for at least part of the year. Many also exhibit at horticultural shows, selling plants directly and taking orders. Such events are often the best way of seeing a wide variety of plants and of obtaining them.

When buying plants, either as bare-root plants by mail order, or in containers, it is much better to do so in spring or summer, when the plants are in active growth. Some grasses are prone to rot if they are divided outside their growing season, or if they are started into growth and then stopped. Divisions obtained in early spring should ideally be kept in a greenhouse or cold frame until early summer, to make sure they are kept growing; a cold wet snap can curtail growth, and the plants may die rapidly. Bamboo should ideally be bought as divisions of young growth, rather than old, thick canes, which will take longer to establish.

PLANTING

Traditional gardening practice lays great emphasis on soil preparation. This may be necessary for many trees and shrubs, but it is often unnecessary for perennials and grasses. A fertile soil, with no obvious problems, needs little preparation. Dig a planting hole twice as wide as the plant, break up the soil at the base with a fork to loosen it, place the plant in the hole, and then backfill it with soil, with any large lumps broken up. The soil level should come as close as possible to the surface of the medium the plant had when in its pot. It is often difficult to tell how deeply to plant bare-root plants, but they often need relatively deep planting to secure them. The ground around a new plant should be firmed with foot pressure to provide good contact between roots and the soil, and to reduce the risk of wind rock.

Soils that are particularly hard or stony should be broken up over a wider area than the planting hole to make it as easy as possible for the young roots to move out into fresh soil. The breaking up of layers of soil below the planting position, by double digging for example, will also help. Modern practice tends to be against the addition of fertilizers or humus to soils when planting ornamental species, because some research indicates that this can sometimes do more harm than good. It is now considered more desirable to encourage plants to develop a healthier root system by making them search for nutrients over a wider area. Large-growing species like silver grass (*Miscanthus*) may well benefit from extra feeding, though.

Bamboo, being less tolerant of difficult conditions than other grasses, needs more care with planting. The addition of plenty of well-rotted manure or homemade compost will enhance the moisture-holding properties of the soil—essential for success with bamboo.

Once planted, most grasses will establish quickly, reaching full height the second year after planting, although it may take several more years before some species bulk out, particularly silver grass (*Miscanthus*). Bamboos are generally much slower to establish; they sometimes take four to eight years after planting before their full height is reached. Plentiful moisture in the first year after planting can assist greatly; these are plants that are very obviously grateful for irrigation.

MAINTENANCE

The only maintenance that most grasses require is the annual removal of dead growth. A few, notably moor grass (*Molinia altissima*) and varieties of *M. caerulea*, such as *M. c. arundinacea* or *M. c. subsp. arundinacea*, tend to collapse in autumn, their seed heads and stems falling, which will necessitate some clearing up. Most others, though, stay firmly upright, not beginning to look ragged until the end of the winter. This is the best time to perform clearing-up operations, cutting back last year's stems and the dead leaves of deciduous species; if this operation is left any later, the emerging young growth may be damaged in the process. Evergreen sedges can also be cut back at this time in order to stimulate the growth of more colorful new leaves.

Those grasses that tend to run need to be looked at at least once a year to make sure they are not spreading too far. Runners can be severed with a spade at the limit of where they are acceptable, and then pulled up. A few grasses can cause a nuisance by seeding prolifically, although it is difficult to predict which will because this is determined by the soil type, light soils often offering a more receptive seed bed. For example, some species of feather grass (*Stipa*) are reported as self seeding by some gardeners, but not at all by others. If seeding does become a problem, the use of a coarse mulch such as wood chips or gravel will greatly reduce it, and any unwanted seedlings that do come through can be pulled out more easily.

BAMBOO FLOWERING

Bamboos, being long-lived evergreens, require little or no maintenance. Fallen leaves should not be removed from the base of the plant, as they provide the best mulch and long-term nutrient. Old clumps can become very dense, in which case you can remove older canes to help improve the appearance of the plant—this is especially beneficial for those that have attractive canes that need to be seen at their best.

Certain problems can arise when the plants flower, which they do very rarely, usually at long intervals of decades or even centuries (some species have never flowered in recorded cultivation). The common belief *all* bamboos die after they flower is wrong, although it happens with some species. Flowering uses energy, and a plant can rejuvenate strongly or die. Badly affected plants will usually produce seed to insure continuity; those that do not mostly recover. Simultaneous flowering of a species occurs worldwide, unless there are clonal variations, proving that all stock of that species originated from a single source. The myth of bamboos dying after flowering grew with the demise of umbrella bamboo (*Fargesia murieliae*), but recent flowerings of other species have proved that regrowth is more usual than death. *Pleioblastus linearis*, *Pseudosasa japonica*, *Phyllostachys flexuosa*, and *Yushania anceps* have all grown strongly after a flowering phase.

ABOVE: *Glyceria maxima* var. *variegata* is a marshland or waterside grass, but it can also be very effective when grown as seen here, in a container sunk into a pond.

LEFT: Silver grass (*Miscanthus*) can be slow to establish but once they are they are very long lived. This is *Miscanthus sinensis* 'Samurai.'

BELOW: Melic (*Melica transsilvanica*) is a drought-tolerant grass, making it especially useful for growing on shallow limestone soils.

PROPAGATION

There are two ways of propagating grasses; by seed and by division. The former method is the best way for obtaining large numbers of plants of natural species. Seed is produced by sexual reproduction, which involves a scrambling of genes, so the offspring are all slightly different. When raising natural species, like Japanese silver grass (*Miscanthus sinensis*), for example, this is no problem, but it is a problem if the plant is a cultivar, a selection whose name is enclosed in single inverted commas, like *Miscanthus sinensis* 'Rotfuchs' or *Carex riparia* 'Variegata.' There is no guarantee the seedlings will have the distinctive characteristics of their parent—seedlings of the former may not have red-tinged seed heads, the latter may not be variegated. Cultivars can only be reliably propagated by division, which produces genetically identical offspring.

SEED

This is the natural way grasses spread themselves over wide areas. Many species in cultivation produce large quantities of seed that is easy to collect. Others do not set seed, or not viable seed. Some, notably bamboo, hardly ever flower and set seed. A limited number of seed companies sell seed of ornamental grasses, while growers of wildflowers for habitat restoration sell bulk quantities of seed of locally native grasses.

The seed of grasses, sedges, and rushes is held in bunches, which are easy to collect, the best time to do so being when the supporting stem turns brown. The seed can usually be separated from the seed head by rubbing it between the fingers. Seed should be stored in paper envelopes in a refrigerator until sowing in the spring. Not only can seed be collected from cultivated plants, but from plants growing wild, too, although this must never be done in nature reserves or from species that are known to be rare.

Most true grasses have seeds that germinate readily in warm moist conditions, but a few, such as species of melic (*Melica*), moor grass (*Molinia*), and silver grass (*Miscanthus*), need a period of warmth, followed by chilling, followed by more warmth, as do sedges and rushes. Ideally, these should be sown in the autumn; if not they should be sown in seed trays in spring, left in a warm place (at around 65°) for four weeks, and then, if germination has not occurred, placed in the refrigerator at around 37° for four weeks. Germination will occur several weeks after removal from the refrigerator.

Gardeners with light soil, with only a limited weed problem, can sow their seed in a nursery bed outside, covering each row of seed with a few millimeters of grit. Most will probably prefer to sow in trays of seed mixture

however. The seed should not be covered with potting medium but with fine grit that allows some light through, which is often needed for germination.

Transplanting should be done before young plants become too enmeshed together in the seed tray or bed. Many are slow growing and delicate, and best pricked out into a nursery bed or containers for the first season. Move them into final positions in late summer or the following spring, when they are in active growth.

DIVISION

All grasses and similar plants form clumps and can thus be propagated by division. Division, like taking cuttings from shrubs or perennials, is a way of reproducing plants that are genetically identical to their parents, so it is ideal for plants with special characteristics that you want to keep. The speed with which a grass forms a clump varies considerably. For example, the bamboo *Chusquea culeou*, which very slowly builds up into a clump, can only be divided after several years' growth, and not surprisingly it is an expensive plant. Running grassess such as ribbon grass (*Phalaris arundinacea* var. *picta* 'Picta') can be divided into several new plants practically as soon as they are bought from the nursery.

True grasses that have a running habit, along with sedges and rushes, can all be divided at any time of year, but given that division is often extremely stressful for the plant, since much of the root system is often severely damaged, it is better to divide it in winter, when water loss is going to be minimal.

However, as a general rule, clump-forming grasses really do not like being divided in winter; several times I have seen nurseries lose their entire new stock when this is done. I myself once lost several hundred needle grass (*Stipa calamagrostis*) when I divided them in spring, only to experience an unseasonal cold snap the next week. In cool temperate climates, early to mid-summer, when the plants are growing vigorously, is the

only safe time. Needless to say, the young plants will need regular watering and possibly shading for several weeks. In regions with a mid-continent climate, where spring is more sudden and predictable, spring is still probably the best time, so plants can establish before the weather gets really hot.

Bamboo is best divided in the growing season, although this is not essential. When dividing, the top growth of bamboo should be shortened considerably, to reduce wind rock and water loss.

Established clumps or large plants may be difficult to divide. The use of two spades, inserted into the clump back to back, to lever the plant apart, is the best way to divide such specimens. Bamboo is incredibly tough—clumps can be hacked apart with a sharp spade or ax, although this can do a lot of damage to shoots. A more precise method is to hammer a spade into the center of the clump, much as you might use a hammer and chisel, or even to use a saw.

Divisions can usually be treated as new plants and put into their permanent positions right away. If the weather is warm and watering widely spaced plants is going to be a problem, or if the divisions are small, it might be better to grow them on in pots or in a nursery bed for a few months first of all—they can then be transplanted later. Transplanting, since little root damage and no shoot damage is caused, can be carried out on any species during the autumn and winter.

ABOVE LEFT AND RIGHT: The varieties of ribbon grass *Phalaris arundinacea* var. *picta* 'Feesey' (left) and *P. a.* var. *picta* 'Picta' are suitable only for places where their running tendencies do not matter.

BELOW: A superb specimen of Japanese silver grass
(*Miscanthus sinensis* 'Giganteus') is bolstered by
terracotta pots filled with fountain grass
(*Pennisetum villosum*), their seed heads spilling
over the sides—a combination that reminds us of
the variety in scale ornamental grasses can offer.

GROWING IN CONTAINERS

Grasses grow well in containers. One reason for this is that they have fibrous superficial rooting systems, it being plants with long, deep thrusting roots that adapt badly to pot culture. Those species that run, producing powerful rhizomes sideways, like some of the *Sasa* bamboos, are not suitable for long-term growing in containers, for the fairly obvious reason that they run out of room; they also make conditions difficult for any other plants sharing the container.

Grasses in containers are best grown in traditional soil-based mixtures, especially taller top-heavy ones, which can so easily overbalance in lightweight soil-free mediums. Soil-based mixtures are also, on the whole, less likely to become soggy and thus initiate root rot during cold wet winter weather. Very tall plants, like bamboos or silver grass (*Miscanthus*), may also benefit from having extra gravel or stones added to the soil for the extra weight they add. Containers for tall plants need to be heavy for the same reason. All containers should have drainage holes in the bottom.

Given the vigorous growth of most grasses, potting mixtures should contain a good supply of nutrients, and for the first growing season, regular fertilizing will be necessary. Any balanced feed is suitable, but ever increasing numbers of gardeners are using the long-life fertilizer granules that have recently become available to the amateur market, which need only be applied once during the growing season. Those based on Osmocote™ are the best, the resin coating being designed to allow more nutrients to be released in warmer conditions, when plants will need more feed.

Frequent watering is vital, especially during hot and dry weather. Indeed, this may well be the main limiting factor in growing bamboo, which is so sensitive to drought, in containers. Bamboo in containers should not be left in the sun, as both the top growth and the roots benefit from cool and shady conditions. Regular watering of bamboo is vital; many years' growth can be permanently damaged by one failure to water.

Given that most grasses are very strong growing, it is probably going to be necessary to remove and divide them every few years; otherwise, the soil in the container may become so packed with roots that the plants are constantly drying out.

INDEX

Entries in **bold** refer to the Directory on pages 78–99. Entries in *italics* refer to captions.

SUPPLIERS

KURT BLUEMEL INC.
2740 Greene Lane, Baldwin, MD 21013
tel: 800 248 PLUG (7584)
fax: 410 557 9785
website: www.bluemel.com

FANCY FRONDS
P.O. Box 1090, Gold Bar, WA 98251
tel: 360 793 1472

FORESTFARM NURSERY
990 Tetherow Road, Williams
OR 97544
tel: 541 846 7269
website: www.forestfarm.com

HERONSWOOD NURSERY
7530 N.E. 288th Street
Kingston, WA 98346
tel: 360 297 4172
website: www.heronswood.com

MORNING GLORY FARM
P.O. Box 423, Fairview, TN 37062
tel: 615 799 0138
fax: 615 799 8864
e-mail: staff@morninggloryfarm.com
website: www.morningloryfarm.com

PLANT DELIGHTS NURSERY
9241 Sauls Road, Raleigh, NC 27603
tel: 919 772 4792
fax: 919 662 0370
e-mail: office@plantdel.com
website: www.plant.del.com

PRAIRE RESTORATIONS INC.
Prairie Creek Farm, P.O. Box 305
Cannon Falls, MN 55009
tel: 507 663 1091
fax: 507 663 1228
website: www.prairieresto.com

TRIPPLE BROOK FARM
37 Middle Road, Southampton
MA 01073
tel: 413 527 4626 fax: 413 527 9853
website: www.tripplebrookfarm.com

TWOMBLY NURSERY
163 Barn Hill Road, Monroe, CT 06468
tel: 203 261 2133 fax: 203 261 9230
e-mail: info@twomblynursery.com
website: www.unusualplants.com

WE-DU NURSERIES
Route 5, Box 724, Marion
NC 28752
tel: 704 738 8300
website: www.we-du.com

WOODLANDERS
1128 Colleton Avenue
Aiken, SC 29801
tel: 803 648 7522

ACKNOWLEDGMENTS

THE AUTHOR would like to thank Andrea Jones for her work in photographing the plants for this book and the owners of all the gardens featured, also Paul Whittaker for this work on the directory and his expert advice on other aspects of the book.

THE PHOTOGRAPHER would like to thank Russell Sharp and Phillip Brown at the Gardens of Portmeirion, North Wales; Lin Randall of the Savill Garden, Windsor Great Park; Angus White of Architectural Plants; Fergus Garrett, head gardener of Great Dixter, for his help and Christopher Lloyd for his permission to photograph; Beth Chatto; Paul and Diana Whittaker from P.W. Plants for their help and hospitaility; Simon Trehane of Trehane Gardens; Marion Holder; Mr. and Mrs. Stoner; Jason Payne; Claudia Beurer and Gerald Singel of Gruga Park Botanic Gardens; M. E Berthelot and M. M Couzet of La Bambouserie; Alan Bloom of Blooms of Bressingham; Sue and Bleddyn Wynn-Jones of Crûg Farm; John Nelson, Tim Smit, and the staff at the Lost Gardens of Heligan; Hugh Angus, Tony Russell and Peter Gregory of Westonbirt Arboretum; Mr. and Mrs. Henry Bungener; Mr. and Mrs. John Lenting; Greg Redwood of the Garden Developments Unit, Royal Botanic Gardens, Kew; Dr. and Mrs. Marsden, Garwell House.

PICTURE CREDITS

l = left r = right t = top b = bottom c = center
Apple Court pp10-11t, p109; Architectural Plants p14c;
La Bambouserie p12t, p21l, p45tr, p47, p51cr, p61cr, p74t, p79, p112t; Terence Black p25tr, p105b; Blooms of Bressingham p31t, p58tr; Beth Chatto p34t, pp52-53b, p55l, p98 No.3; Cowley Manor pp76-77t; Garwell House p57; Great Dixter p60b, p63, p64tr, p105tr; Gruga Park p13c, p30 tr, p51r, p59tl, p65, p73b, p104; Gillian Harris p70tr; Hermanshof p15l&r, p24, p62l&r, p64bl, p107b, p112b; C. Jannsen p20, pp42-43b, p55tr, p67l; Japanese Gardens p19b, p22l, p29tr, p33t, p46, p56b, p59tr, p66t, p68l, p72, p73trl&c, p74b; Carol Klein, Chelsea Flower Show p48; Lady Farm p3b, p9, p31b, p61tr, p110; Piet Oudolf p3t, p12c, p43b, p53t, p61tl, p75bl, pp76-77c&b, p111; P.W. Plants p6, p14br, p28-29t, p35tl, p49, p50t&bl, p67r, p70tl&70b, p71, p96 No. 1; Rowden Gardens p35tr, p64tl; Royal Botanic Gardens, Kew p10c, p58tl, p64br, p106b; Scottish Bamboo Nursery p38t.

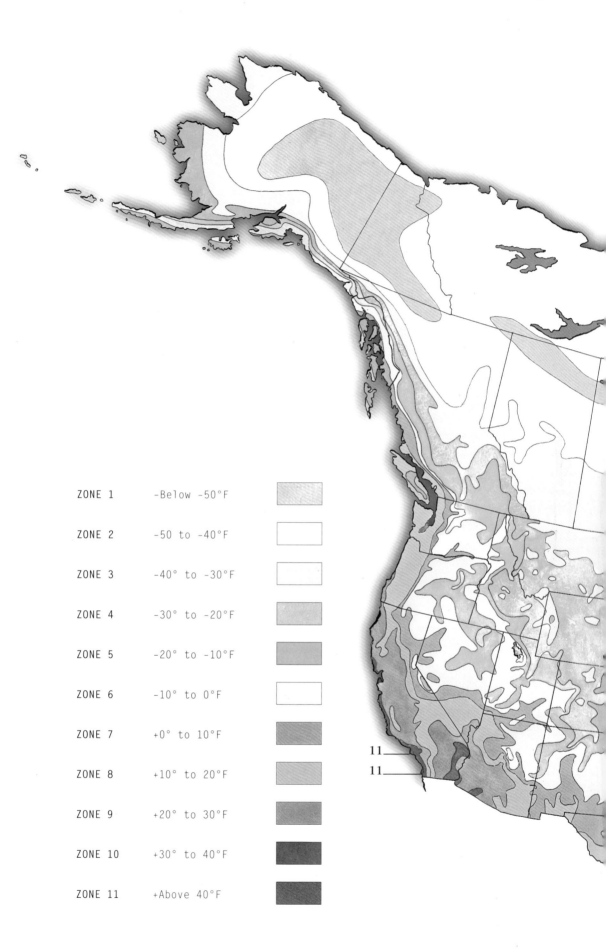

ZONE 1	-Below -50°F
ZONE 2	-50 to -40°F
ZONE 3	-40° to -30°F
ZONE 4	-30° to -20°F
ZONE 5	-20° to -10°F
ZONE 6	-10° to 0°F
ZONE 7	+0° to 10°F
ZONE 8	+10° to 20°F
ZONE 9	+20° to 30°F
ZONE 10	+30° to 40°F
ZONE 11	+Above 40°F